The Tunnel under the Lake

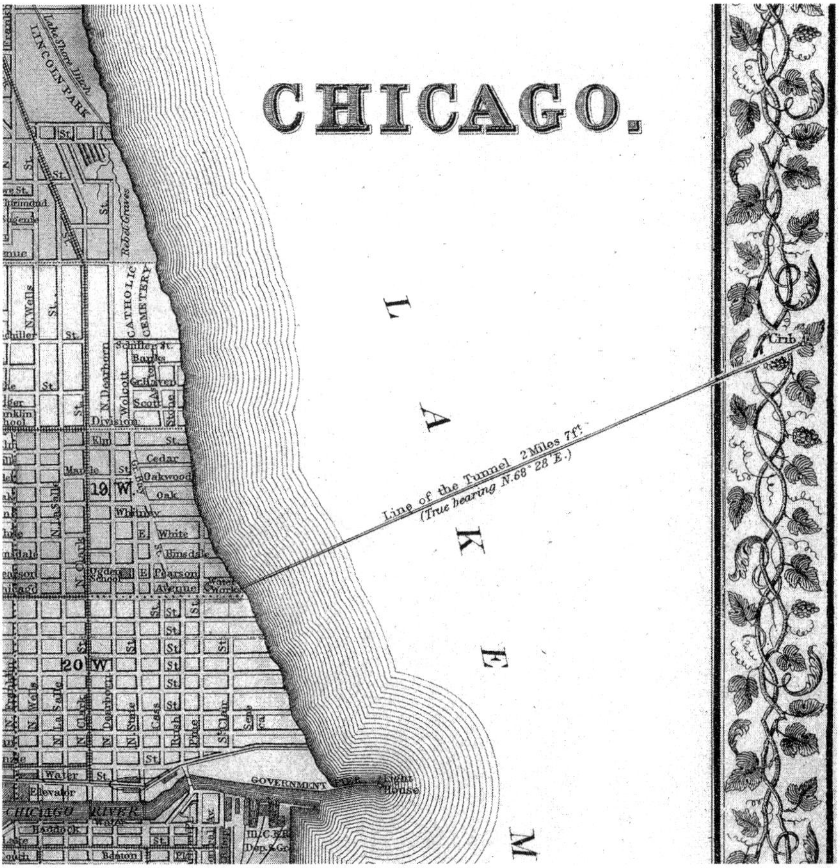
CHICAGO.
LAKE M
Line of the Tunnel 2 Miles 7ft.
(True bearing N.68° 28' E.)
Crib
LINCOLN PARK
Lake Shore Ditch
Rebel Graves
CATHOLIC CEMETERY
GOVERNMENT PIER
Light House
CHICAGO RIVER
Elevator
19 W
20 W
Division
Elm
Cedar
Oakwood
Oak
White
Pearson
Water
Schiller St.
N. Wells
N. Dearborn
N. Clark
N. La Salle
N. State
Wolcott
Rush
Pine
St. Clair
Ill. C. R.R. Dep. & Gr.

The Tunnel under the Lake

The Engineering Marvel That Saved Chicago

Benjamin Sells

NORTHWESTERN UNIVERSITY PRESS
EVANSTON, ILLINOIS

Northwestern University Press
www.nupress.northwestern.edu

Cover and frontispiece: Map of Chicago, 1870, showing the great Chicago lake tunnel and the Two-Mile Crib. From Samuel Augustus Mitchell, *Mitchell's New General Atlas* (Philadelphia: S. A. Mitchell Jr., 1870). Courtesy David Rumsey Map Collection, www.davidrumsey.com.

Printed in the United States of America

10 9 8 7 6 5 4 3 2 1

Library of Congress Cataloging-in-Publication Data

Names: Sells, Benjamin, author.
Title: The tunnel under the lake : the engineering marvel that saved Chicago / Benjamin Sells.
Other titles: Second to none: Chicago stories.
Description: Evanston, Illinois : Northwestern University Press, 2017. | Series: Second to none: Chicago stories | Includes bibliographical references and index.
Identifiers: LCCN 2017000235 | ISBN 9780810134744 (pbk. : alk. paper) | ISBN 9780810134751 (e-book)
Subjects: LCSH: Municipal water supply—Illinois—Chicago—History—19th century. | Water-supply engineering—Illinois—Chicago—History—19th century. | Water tunnels—Illinois—Chicago—History—19th century. | Waterworks—Illinois—Chicago—History—19th century. | Chesbrough, Ellis Sylvester, 1813–1886. | Chicago (Ill.)—Buildings, structures, etc.—History—19th century.
Classification: LCC TD225.C5 S45 2017 | DDC 363.61097731109034—dc23
LC record available at https://lccn.loc.gov/2017000235

CONTENTS

The tunnel itself will be, when completed, one of the wonders of the world, worthy of taking its place as the eighth. We believe that such a work has never before been undertaken. Aqueducts have ere now been carried across valleys, and canals have been driven through mountains, but not on the airtight principle. It remains for Chicago, herself a wonder of the world, to show to its civilized portion how a watertight canal can be formed under the bed of a mighty lake, running out to a distance of two miles towards the rising sun. . . . Chicago is going ahead, and her every effort is a gigantic stride towards perfection.

—*Chicago Tribune*, July 25, 1865

INTRODUCTION

The great Chicago lake tunnel was a 2-mile-long underwater tunnel, dug and bricked by hand 35 feet below the floor of Lake Michigan, to a massive offshore "crib" that protected the tunnel's intake. The tunnel's purpose was to bring fresh water to a city in dire need. It was one of the most significant engineering feats of the nineteenth century, and at the time of its opening, in March 1867, it was hailed by Mayor John B. Rice as "the wonder of America and of the world." For once, the actual feat matched the political rhetoric, and it is not too much to say that without the Chicago lake tunnel we would not have the Chicago we know today.

The story of the great Chicago lake tunnel is one of bigger-than-life individuals, of exceptional skill and imagination, and of human toil almost beyond comprehension. It is the story of a life-and-death struggle against devastating diseases of then unknown origin that threatened a city in its infancy, a struggle waged even as the nation was torn by civil war. And it is the story of one remarkable engineer whose vision, perseverance, and practical genius would save Chicago from itself.

In these days of rapidly changing technology, it can be easy to overlook the impact of technologies so ingrained in our daily lives that we take them for granted. We flip a switch, and light brightens the night. We turn a key, and a motor revs to life, propelling us at incredible speeds across roadways that span a continent. We turn a handle, and clean water streams into a glass. Although it is difficult to imagine, we could do without electricity or engines and the benefits they provide. But without clean water we would perish. Without water there *is* no we—it is simply impossible for people to endure, or for civilization to proceed, without a supply of clean water. Bring clean water to the most desolate of locations, and life will surely follow. Deprive the most advanced location of water, and soon there will be nothing.

There are few better examples of the transformative, life-giving power of clean water than the results of discoveries and transformations that took place in the Chicago of the mid-nineteenth century. In 1833, Chicago was just a hamlet on the mudflats and marshes surrounding the Chicago River. Like the Native Americans who had lived there long before, early settlers drew their water directly from the river and the lake. As the settlements grew, families began to look for water sources closer to hand and within their own control. The topography of the area made digging private wells an easy task.

The area along the river was exceptionally flat, with the riverbanks only about 3 feet above the water's surface. The sandy soil and high water table meant that people had to dig only 6 to 12 feet to find water.

Along with their need for water, the settlers had to find ways of disposing of their waste. The same factors that made digging wells easy also made it easy for families to dig privy vaults. But this ease was accompanied by unforeseen risks. Below the sandy soil lay an impervious layer of clay. Given the proximity of the wells to the privy vaults, contaminants from leaking vaults would permeate down to the clay and then flow laterally to find their way into the water supply, creating a persistent, and deadly, problem.

It did not take long for individual efforts to provide pure water to begin to falter. In a town on its way to greatness, individual water supplies quickly met their limits as collective needs took precedence. The first public well was dug in 1834. In the 1840s, the recently incorporated city chartered a private company to construct a network for the supply and distribution of water. Not until 1851 did the city itself take over responsibility for supplying fresh water to its residents.

To appreciate how the need for fresh water quickly outstripped individual and private efforts, we need only look to the unprecedented growth of Chicago in the1800s. Chicago was incorporated as a town in 1833, with a population of 350. Within a year, that number grew to 2,000. By the time Chicago was incorporated as a city, in 1837, its population had doubled to 4,180, and by 1846 it had reached 14,169. By 1860 the city numbered 109,206 residents, and by 1870 that number had tripled to 306,605. (For further documentation of Chicago's yearly population growth during this period, see appendix A).

In 1880 the city's population reached 503,000, and by 1890 it had doubled again to over 1 million residents, with Chicago surpassing Philadelphia to become the nation's "second city." A rapidly expanding population was typical of America's growing cities in the nineteenth century—between 1840 and 1880, Detroit's population expanded 13 times over, from 9,000 to 116,000, and the population of St. Louis expanded 22 times over, from 16,000 to 351,000—but Chicago's population, during the same years, multiplied *126 times* over, from 4,000 to just over 500,000 residents.[1] By any measure, Chicago was in a different league.

From the very beginning, the story of Chicago would be intertwined with the city's quest for pure water. There was never enough, and Chicago was forever playing catch-up. In 1858, for example, the superintendent of the Board of Water Commissioners, noting that the city's growth had defied all prior projections, concluded, "There is very little danger of our being likely to extend the [water]works faster than the wants of the city will require."[2] In one of the great ironies of early America, a city poised on the banks of a seemingly inexhaustible supply of fresh water would increasingly find itself wanting, a victim of its own exponential success.[3]

In the early days of private wells and privy vaults, families suffered from contaminated water because they were not able to protect their drinking water from their own waste. As the city grew, this problem would magnify and extend beyond the contamination of private wells to include the natural waterways that had made the city possible. Although the Chicago River and Lake Michigan provided a readily accessible supply of fresh water, the same sources quickly became the common receptacles of the city's waste. The enduring dilemma of Chicago would be how to accommodate these conflicting uses of water. Solving this dilemma was not merely a question of convenience; it carried the most deadly of consequences. From the early 1830s on, the city suffered repeated deadly outbreaks of cholera due to contaminated water. In 1854 alone, 2 percent of the city's population was killed as the city endured the horrifying ordeal of one of the worst cholera outbreaks in American history.[4]

* * *

The story of Chicago's quest for pure water begins long before the arrival of the area's earliest human residents. This a story told on a timeless stage, a story of ancient seas teeming with life that created the literal bedrock for our own life, of towering glaciers that sculpted the basins that became the Great Lakes, and of raging waters that carved the rivers that would someday carry the canoes launched by our insatiable curiosity. It would take millennia for nature to prepare a unique spot on the southwest shore of Lake Michigan that would become the fulcrum of a nation.

There was a time when the area now home to the Great Lakes was a great tropical sea. The calcium carbonate deposits from the shells and life forms of that sea eventually became the limestone bedrock of the region. In time the sea was replaced by a swampy delta that in turn, during the Ice Age, would itself become covered and compacted with glaciers that were at places several miles thick. As the glaciers receded, the lakes began to settle into their respective basins. But even as they settled, the lake waters never stopped seeking ways to return to the sea.[5]

On the west side of the ancient lake that would become Lake Michigan, an outlet formed, allowing water to flow by way of a trough gouged through the solid limestone into the Mississippi Valley and then to the Gulf of Mexico. The trough began, roughly speaking, at present-day Summit, Illinois, about 11 miles west of Chicago, and ended 20 miles southwest, at present-day Lockport. The bottom of this trough was about 6 feet above the current level of the lake and was 1 to 2 miles wide.[6]

As the shores of the ancient lake receded, a ridge of limestone rock running more or less parallel to the western shore of the current Lake Michigan arose to separate two great drainage basins. To the west of this divide, water flowed down the Des Plaines and Kankakee Rivers to the Illinois River and then to the Mississippi River and the Gulf of Mexico. To the east, water combined with the water of the other Great Lakes and flowed through the St. Lawrence River to the Atlantic Ocean. With the appearance of this subcontinental divide (known as the St. Lawrence Divide), waters that had once flowed through the great trough westward into the Mississippi Valley began to drain eastward instead into Lake Michigan through what would become the Chicago River.

Between the west fork of the south branch of the Chicago River and the Des Plaines River lay a marshy area that would come to be called Mud Lake. At its westward end was a portage, well known to the Native American inhabitants, that provided a critical link between the Des Plaines River to the west and the Chicago River to the east, effectively making a bridge between Lake Michigan and the Gulf of Mexico. This connection came to be known as the Chicago Portage.

During their explorations of the region in 1673, the famed explorers Louis Joliet and Jacques Marquette were shown the portage by the local inhabitants, and Joliet became the first European to recognize the potential for creating a waterway at the portage connecting Lake Michigan and the Gulf of Mexico. Joliet gave a verbal account of his journey and its discoveries to Father Claude Dablon, who in turn, in a letter dated August 1, 1674, recounted Joliet's prescient vision for this waterway, a vision that would take almost two centuries to become reality. According to Father Dablon, Joliet gave the following report:

> We can quite easily go to Florida in boats, and by a very good navigation. There would be but one canal to make, by cutting only one-half a league of prairie, to pass from the Lake of the Illinois [Lake Michigan] into St. Louis river [Illinois River]. The route to be taken is this: the bark should be built on Lake Erie, which is near Lake Ontario; it would pass easily from Lake Erie to Lake Huron, from which it would enter the Lake of the Illinois. At the extremity of this lake would be the cut or canal of which I have spoken, to have a passage to St. Louis river, which empties into the Mississippi. The bark having entered this river, could easily sail to the Gulf of Mexico.[7]

Over the next hundred years, as various parties struggled for control of the entire area, the idea of a canal re-creating the ancient waterway from Lake Michigan to the Mississippi River was set aside. On August 3, 1795, a peace treaty was signed whereby the Native American inhabitants (the Potawatomi) ceded ownership of fourteen parcels of land to the United States, including one parcel 6 miles square at the mouth of the Chicago River. The treaty specifically mentioned that the people of the United States were to have free passage from the mouth of the Chicago River to "the commencement of the portage between that river and the Illinois and down the Illinois River to the Mississippi."[8] This was the first official acknowledgment of the portage connecting Lake Michigan with the Mississippi River.

It is difficult to overstate the importance of the Chicago Portage to the development of Chicago, the American West, and indeed America itself. Its pivotal importance as a trade route was a main factor in bringing the first European settlers to the area, and it would later connect the markets of the East with the products of the West. It would provide the natural template for the Illinois and Michigan Canal and, even later, for the Sanitary and Ship Canal, which would finally restore the ancient flow of the Chicago River westward. It was this saddle—just feet above the level of Lake Michigan, and carved over millennia by waters seeking their way to the Gulf of Mexico—that would help transform a small settlement into a great metropolis. And it was the interplay between the area's geological makeup and the nature of its waterways that would provide the evolving backdrop to Chicago's quest for pure water.

* * *

Chicago's geological past had created a natural connection between the Gulf of Mexico and the Atlantic Ocean. All that was needed now was a reason to make the trip between those two bodies of water. By the 1800s, the expanding nation was more than happy to oblige. Chicago, situated between the growing industrial areas of the Northeast and the vast frontiers of the West, was poised to take full advantage of the technologies of its day. The first telegraph line reached the city in January 1848, and on April 10 of that year cheering citizens welcomed the *General Fry* as the first canal boat to arrive in Chicago from Lockport, towed up the Illinois and Michigan Canal by the steam tug *A. Rossiter*. The canal formally opened on April 16, and on April 23 the canal boat *General Thornton* arrived from New Orleans with a load of sugar bound for Buffalo. The sugar arrived in Buffalo on April 30, two weeks ahead of consignments sent to the same destination by ocean steamship traveling first to New York and then up the Hudson River and the Erie Canal.[9] The significance of the Illinois and Michigan Canal as a trade route of national importance was recognized on June 20, 1848, in a speech on the floor of Congress by a young Abraham Lincoln.[10] Also in 1848, the first railcar arrived in Chicago, and the first 11 miles of track were laid west of the city. Chicago had quickly become the water transportation hub of America, and within ten years it would also be the center of a rail network with over 3,000 miles of track.[11]

As the new heartbeat of America, Chicago attracted many of the leading businesspeople and industrialists of the day. Speculation, the coin of the realm, helped fuel Chicago's explosive growth in both population and industry. But by the 1850s the city was outgrowing itself, and increasingly its greatest threats came from within. Built too fast, the city had been left waist-high in filth. The river, the city's life-giving artery, was already clogged with refuse, sewage, and other horrors.

Having effectively eliminated the river as a source of clean water, the city turned to the lake. Inlet pipes were run offshore in an attempt to reach beyond the polluting effects of the river, but the lake proved a tough customer. The city was able to extend its drinking straw only a few hundred feet offshore, and the water obtained in this way was often muddy. Even worse, the relatively warm water of the shallow inlet attracted thousands of small fish that found their way into the water supply. It was not uncommon to open a faucet and have a wriggling minnow end up in one's glass.

Water and waste, waste and water—the two were inextricably linked, and together they threatened to undermine the growing city in the most basic ways possible. Without a way to dispose of its waste and provide clean, safe drinking water to its citizens, the city could not endure. The challenge had become existential, and patchwork solutions no longer sufficed. For the young city, it was time to do or die.

The task of solving Chicago's great dilemma fell to a forty-two-year-old engineer, Ellis Sylvester Chesbrough. In 1855, after having served as Boston's first city engineer, Chesbrough was hired as chief engineer of Chicago's newly formed Board of Sewerage Commissioners. It was a hire that changed Chicago forever. Chesbrough would become one of the most important, if unsung, people in the history of Chicago. Although politicians and entrepreneurs of the day sought and often found the limelight, Chesbrough was content to do his indispensable work in the waste and mud choking the city. He would design and implement the first systematic sewerage system in America, a system that literally raised Chicago from the muck and revolutionized sanitation management for the entire country. That success alone would have been enough to secure his place as one of the great engineers of the nineteenth century. But his city needed more from him. It needed fresh water, and it needed it badly.

Chesbrough had only to cast his eyes eastward for the solution. The clean water was right out there, tantalizingly close to the lips of a thirsty city. The question was how to get it to the waiting faucets and hydrants. As we have seen, experience had shown that running inlet pipes into the tumultuous lake was a solution difficult to implement and sustain, and that the water acquired from so close to shore was of inferior quality.

Chesbrough's answer was a tunnel. Not just any tunnel, but a tunnel under the lake, 35 feet below the lakebed, dug and bricked by hand and reaching 2 miles offshore to an intake housed by a massive crib of a size and kind never before constructed. Here the water ran cool and clear, beyond the reach of the deadly contamination of the river and the lakeshore, beyond the mud and the minnows. This was the Chesbrough Plan, a plan of audacious simplicity. Many lesser minds would argue that the feat could not be done, but Chesbrough would prove to be a man not easily daunted by doubt.

The following pages tell the story of the conception of the Chesbrough Plan and the planning and building of the Chicago lake tunnel and crib. We will begin with Chicago's early days, when fresh water came by way of buckets, wells, and horse-drawn water carts. We will then examine the first collective efforts to provide for the water supply by way of private companies, efforts that in turn led to a significant historical shift whereby the city would take over responsibility for providing water to its residents. We will see how persistent problems with water quality, and the devastating cholera outbreaks that ravaged the city, drove the demand for a more stable and secure supply.

Chicago's rise as the city of the century simply cannot be understood without an appreciation of its desperate quest to rid itself of its waste while also finding a source of pure water. Water made Chicago, but water also came to be the city's greatest threat. Growth was Chicago's hallmark, but growth also came to be the city's greatest adversary. The story of how Chicago met these challenges is as much about the human spirit as about engineering prowess. The fact that Chesbrough was able to convince the city to take what many saw as an unprecedented risk is in some ways as remarkable as the project itself. We will look closely at the varying interests and parties involved, and we will see how an engineer from Boston managed to navigate the intricacies of Chicago politics.

We will then examine in detail the actual construction of the Chicago lake tunnel and crib. To say that these projects were engineering marvels hardly does them justice. In a time when large engineering projects were reshaping the nation, the lake tunnel and crib captured the national imagination in a way that canals and aqueducts did not. Every day seemed to bring a new challenge that required a new solution. The sinking of the land and lake shafts, with their iron cylinders so large that they had to be moved in railcars of their own; the unique masonry demands of bricking the tunnel; the task of figuring out how to dig a straight line for 2 miles underground, where compasses were useless; the risk of explosions and fires in a tunnel beneath the waves; workers' constant fear of a catastrophic, inescapable collapse and a watery grave; the urgency of finding ways to sink and stabilize a crib the size of a courthouse in the churning waters of Lake Michigan; even the need to deal with human foibles, from greed to murder to insanity—these were the daily challenges of the Chesbrough Plan. Meanwhile, back on shore, politicians and naysayers kept up such a constant chorus of negativity that they almost seemed to hope the project would fail.

But it did not fail. Almost three years after the first spade of dirt was turned, the final stone for the tunnel was pushed into place. A few months later a great celebration was at hand, an entire city rejoicing at the success of a project so boldly conceived and flawlessly accomplished. The inlet gates were opened and the mighty pumps engaged, and when the first rush of clean water made its way through the tunnel to sate a waiting city, there was little doubt that the great Chicago lake tunnel was indeed, in the words of Mayor Rice, "the wonder of America and of the world."

Chapter 1

Buckets, Wells, and Water Carts

The year 1832 was an auspicious if also dark one in the history of both Chicago and the United States. At the time, Chicago was an unincorporated hamlet of only a few families living alongside the Chicago River. The most significant building was Fort Dearborn, a major military outpost on the south bank of the Chicago River that had been built to guard the already important port of Chicago. As the year dawned, no one could have envisioned what was to come or the impact that the year's events would have on the development of Chicago and on its long and arduous quest for pure water.

The Black Hawk War began in May of that year, and the federal government, fearing an attack on Fort Dearborn, decided to send reinforcements to the area. In June, President Andrew Jackson appointed General Winfield Scott to take command of the war, and Scott gathered about 950 troops from along the eastern seaboard to head west to Fort Dearborn. Two steamboats were chartered to transport the troops, and in late June the steamer *Sheldon Thompson* left from Buffalo, New York, with General Scott and his staff, six companies of artillery, and two companies of infantry. Shortly thereafter, on July 3, the steamer *Henry Clay* followed with seven companies of infantry.[1]

It was a cursed voyage. The *Henry Clay* was barely under way when the troops began to fall sick. The stricken ship landed on July 7 near Fort Gratiot and the St. Clair River. Frightened soldiers fled the vessel, and many of them were later found dead in the surrounding countryside. By July 9, of the seven companies of infantry that had set sail just days before, only 68 men remained alive.

The *Sheldon Thompson*, meanwhile, had largely escaped the first days of virulence. But she would not escape unscathed. By the time the ship arrived in Chicago, on July 10, 76 troops had fallen sick and 19 had died. In all, of the original 950 men who had left Buffalo, only 350 made it to Chicago; disease and desertion took the rest.

The onboard killer was cholera, a scourge that would haunt and ravage the growing city, indeed the entire country, for decades to come. The ill-fated timing of Scott's voyage had coincided with the disease's arrival in the United States as one manifestation of a pandemic that was to take hundreds of thousands of lives around the globe. In New York alone, where the disease had arrived from Canada in June, over 3,500 people died out of a population of about 200,000.

It was a horrible disease, made even more fearsome by the severity and speed of its onset. People who woke up healthy were suddenly besieged with vomiting, diarrhea, and wrenching cramps and were dead by sunset. Severe dehydration often left the skin of the afflicted a deathly bluish-gray. So terrifying was the onslaught of cholera that thousands of people took flight from affected areas, hoping to outrun its reach. Some made it. Others, many, did not. Even worse, the doomed ones carried the disease with them in their flight, thus spreading the contagion to smaller towns and outposts, which in turn quickly succumbed.

When Scott's troops arrived at Fort Dearborn, the garrison, swelled by interest in the Black Hawk War, numbered about 1,000. The medical officer of the fort would later recount the "frightful havoc" of the disease, which in the end infected 1 in 5 and killed over 60.[2] General Scott did his best to quarantine the fort, but the illness spread inexorably to the surrounding countryside. Of the troops that had originally set out from Buffalo to assist in the Black Hawk War, few ever made it to their appointed task.

Early on a connection was made between contaminated water and cholera, although the actual mechanism of transmittal was unknown. The discovery of bacteria and their role in disease was still decades away. The reigning theory in the early nineteenth century was that disease was carried and spread by "miasma," or noxious vapors, also called "death fogs" or "night air." The belief was that the miasma contained disease-carrying particles arising from filth, excrement, and decaying organic matter. The filth theory of disease basically said that if something smelled bad, it was bad for you. The response, then, was to try to get rid of the filth that gave rise to the bad smell.

This theory carried over to water.[3] By the sanitation standards of the early nineteenth century, if water looked and smelled clean, it was assumed to be so. Conversely, foul-smelling water was to be feared, and for many years this fear drove much of Chicago's effort concerning sewerage and the water supply. Not until the 1880s would the filth theory be definitively replaced by the germ theory of disease.

In Chicago's earliest days, settlers had limited choices for obtaining water. Getting water from the river and the lake was straightforward but also cumbersome. Many early settlers simply took their pails to the closest source and collected water for their daily needs. Of course, that was an option only for people who lived close enough to the river or the lake. Other options included collecting rainwater, which had the benefit of purity and softness, and, in the winter, collecting melting ice and snow. As Chicago grew, however, the river and the easily accessible lake waters became contaminated with sewage and refuse from the city's industries. Similarly, smoke from homes and industries contaminated winter ice and snow and gave the water obtained from that source a creosote taste.

Figure 1. Private well. Source: Jennie Hall, *The Story of Chicago* (Chicago: Rand McNally, 1911).

As it became more inconvenient, and undesirable, to get water directly from the river and the lake, private wells became ubiquitous. The natural geological conditions encouraged the use of private wells. The original level of the settlement was nowhere more than 12 feet above the level of the lake, and the riverbanks were only 2 to 3 feet above the surface of the river. The ground itself was a bed of sand left behind by the receding waters of the ancient lake that eventually became Lake Michigan. Below that was blue clay, which extended down about 100 feet to the underlying limestone bedrock. The soft surface sand, and the fact that one rarely had to dig more than 6 to 12 feet to hit water, made digging wells relatively easy.

Along with the need for clean water came the problem of how to properly dispose of black water (sewage) and gray water (household wastewater from cleaning, bathing, and the like). The same factors that made digging wells easy also made it easy to dig privy vaults to collect black water. Often, however, the vaults were dug too close to the wells that had been dug for a water supply. The vaults of the era were notorious for leaking, and contaminants from the privy vaults permeated the sand until they encountered the impermeable layer of clay, at which point their downward progress was halted and they flowed horizontally, finding their way into the wells. This problem was compounded by the habit of throwing garbage and gray water onto the ground so that those contaminants also found their way into the wells. As one early resident recounted, water drawn from the wells began to taste "a little brackish at first, then saltish, and finally it had a perceptible odor, which ultimately became offensive."[4]

Before long, as Chicago's population exploded and its accessible waterways were fouled by the city's waste, the problems associated with contaminated water extended beyond private wells and privy vaults. How to dispose of waste while maintaining a fresh water supply became a defining question for Chicago's growth and development.[5]

Chicago was incorporated as a town in 1833, and on November 7 of that year, with the specter of cholera hanging in every foul odor, one of the first ordinances passed by the town council was directed toward protecting the water supply. It was thereafter illegal to throw dead animals into the river, and violators were subject to a fine of $3. There soon followed ordinances forbidding the dumping of garbage into the river or onto vacant land, all in an effort to eliminate the dreaded death fogs.[6] In 1834 the first board of health was established, and an ordinance was passed requiring every male over the age of twenty-one to assist in cleaning the streets, with a fine of $5 for noncompliance (by September 1834, when the cholera outbreak had subsided, this fine was reduced to $1).[7]

As it became more and more difficult to obtain clean water from private wells and the river, local entrepreneurs came up with a new supply. In the 1830s and early 1840s, horse-drawn water carts supplied water to residents who preferred lake water over the more problematic well water and river water. Each cart held a large cask, and the watermen would back their carts down into the lake at the foot of Van Buren Street to get their fill. Then it was off to make their rounds, selling water to local residents at 5 to 10 cents a barrel. Originally water was dipped from the casks to fill customers' barrels, but later the water was dispensed through a 4-inch leather hose attached near the bottom of the cask. The water carts did a thriving business, given the impurities of the water obtained from private wells, and the inconvenience of trekking back and forth with pails to the river or the lake. Companies were organized and contracts were signed for routine delivery of water to residents and businesses. The water carts could be seen plying their trade into the 1850s.[8]

Despite their industry, however, the watermen with their carts were no match for the exponentially growing demand for fresh water. The often tempestuous lake could make filling the casks impossible, and in periods of bad weather the water drawn from the lake was muddy with stirred-up sediment. Residents called out for options, and on November 10, 1834, just one year after Chicago was incorporated as a town, the city made its first foray into providing a public water supply. The town trustees authorized spending $95.50 to dig a public well at the corner of Cass Street (now Wabash Avenue)

Figure 2. Water supply of early Chicago. Source: Chicago History Museum, ICHi-85646. W. E. S. Trowbridge, illustrator.

and Michigan Street (now Hubbard Street). Unfortunately, there was at that time no distribution system apart from the practice of people coming to the well to fill their own buckets and barrels, and so the public well was of real benefit only to people living nearby. Other Chicagoans still had to get their water from wells, the river, the lake, or the water carts.[9]

By the end of 1835, Chicago was booming. The first census of Chicago, completed in 1835, showed "3,265 persons, 398 dwellings, 4 warehouses, 29 dry goods stores, 19 grocery and provision stores, 5 hardware stores, 3 drug stores, 19 taverns, 26 groceries (probably liquor stores) and 17 lawyer's offices."[10] This was an almost tenfold increase over what had been reported barely three years earlier. Already the local population had endured the first decimating onslaughts of cholera. This fact, coupled with the inadequacy of the existing water supplies and the impracticality of using them, made it readily apparent that the days of wells and water carts were coming to an end. A more formal and systematic way had to be found to supply water to the growing population.

On January 18, 1836, the state legislature, by a special act, incorporated the Chicago Hydraulic Company. Curiously, the state legislature acted without any reference to the board of trustees or the residents of the town. There is no evidence that the trustees or residents were opposed to the plan, but the legislature's act of granting a private company access to public streets, and giving the company the right to use them, clearly implied the legislature's belief that such power belonged to the state legislature and not to the town.[11]

The Chicago Hydraulic Company was given a seventy-year charter, and its capital stock was limited to $250,000. It was organized on March 19, 1836, with George W. Dole as president; Gurdon S. Hubbard, David Hunter, Gholson Kercheval, and William Forsythe as directors; and Edward W. Casey as secretary (other incorporators included James H. Campbell, R. A. Kinzie, and Solomon Wells). The original charter gave the newly formed company four years from the passage of the act to begin construction of a waterworks to supply the town with pure water.[12]

A severe financial crisis hit the United States in 1837, temporarily derailing the Chicago Hydraulic Company's efforts to begin construction. It was not until 1840, when its four-year window was beginning to close, that the company finally began construction. Ira Miltimore was named machinist and superintendent of the work, and he began construction in 1840 of a reservoir at the corner of Lake Street and Michigan Avenue, south of the river.[13]

The hiatus between the passage of the enabling charter and the beginning of construction led some residents to question whether the water supply should indeed be provided by a private entity. The City of Chicago had been officially incorporated on March 4, 1837, and many citizens believed, contrary to the legislature's assumption of its power to grant a private corporation access to and use of city property, that the city itself had the corporate authority to construct and run a waterworks under its own newly vested powers. In 1839, the *Chicago Daily American* warned against extending monopoly powers for supplying the city with water to any private company and expressed the hope that the city would soon turn its attention to providing water for its residents.[14] The paper's view presaged the city's eventual takeover of the water supply.[15]

After a fitful start, the new waterworks finally became operational in 1842. Residents assembled in great anticipation of the fateful moment when the trusty 25-horsepower pump would be fired up to provide the first public drink. It was a "gala day," and as the first trickle grew to a steady stream, "a great cheer went up."[16] A new day had arrived in Chicago's quest for pure water.

By mid-1842, the Chicago Hydraulic Company had spent about $24,000 on the waterworks. A two-story brick building had been erected, with a pier running into the lake. The steam engine powered a double-action pump that had a working barrel 14 inches in diameter and a 44-inch stroke. Varying accounts described the suction pipe by which water was drawn from the lake as being between 14 and 18 inches in diameter, and between 320 and 700 feet long. The inlet pipe was laid on the top of the pier, with an elbow at the lake end that dipped down into the lake 5 or 6 feet. Water was pumped into two reservoirs whose capacity, by varying accounts, was said to be between 500 and 1,250 barrels. The distribution pipes were wooden logs bored out by hand, 5 inches for the main lines and 3 inches for the subordinate ones, and reinforced with iron hoops. (These early wooden pipes proved so durable that decades later they were still occasionally dug up in the process of other work and were found to be intact.) By June about 2 miles of pipe had been laid.[17]

Although the new waterworks was a significant step forward, it initially served only the south side of Chicago and, later, a small portion of the west side. (From Chicago's earliest days, the river provided the defining nomenclature for the city. The main stem, running east to west, formed a T with the north and south branches. The north side and the south side of the city lay north and south, respectively, of the main stem, and the west side of the city was west of the axis formed by the north and south branches.) On March 1, 1845, the state legislature passed an act granting a charter to a new company, the Lake Michigan Hydraulic Company, to supply water to the city's north side. This action was again taken without any reference to the now formerly vested city government or its citizens. The initial incorporators were Thomas Dyer, William B. Ogden (Chicago's first mayor), William E. Ives, Walter L. Newberry, Henry Smith, Mahlon D. Ogden, John H. Kinzie, and John B. Turner. The act was not to take effect if the Chicago Hydraulic Company managed to begin building the works necessary to supply the north side within six months. In the end, the Chicago Hydraulic Company never did begin such construction, but neither did the Lake Michigan Hydraulic Company ever proceed with providing any water for the city.[18]

In 1842, James Long entered into an agreement with the Chicago Hydraulic Company to oversee and maintain the company's pumping system for ten years, at no cost, in exchange for the right to use the steam engine's excess power to run a flour mill. To this end, he constructed the Hydraulic Mills at the corner of Lake Street and Michigan Avenue, near the lake. The agreement worked well for Long but was not without its hardships. In winter, for example, the pipes would become disarranged by the heaving of frost, and Long frequently had to spend hours throwing water on them to freeze up cracks before the pumps could be engaged. He also had to deal with fluctuations in the lake levels that sometimes left the end of the intake pipe out of water.[19] Despite such hardships, however, the mill did a good business, and by 1848 it was producing fifty barrels of flour per day.[20] It continued to thrive until 1853, when the city took over the waterworks.

The citizens were notably proud of their waterworks, and it is indisputable that the purity of its water supply was superior to that of the water available from private wells or from the river. This purity, however, did not come cheap. Costs ranged from $10 a year for families of no more than five persons to between $15 and $125 a year for boardinghouses (provided no baths or bathhouses were connected), and from $50 to $200 a year for taverns and hotels. At these rates, only a few hundred families could afford the new water supply.[21]

Despite its comparatively superior quality, the water provided by the waterworks became muddy whenever the lake was tossed by storms or high winds. Even worse was what came to be called the "fish nuisance." The inlet pipe for the water supply was in shallow water, and with the onset of cold weather, thousands of tiny fish would congregate in the warmer water around the mouth of the inlet, where they promptly got sucked into the water supply. Needless to say, this caused havoc with the pump, but many fish also found their way into the distribution pipes, only to come wriggling out of faucets, to the dismay of the homemakers of the day. Bits of wood and other debris also routinely made it into the homes and businesses supplied by the system.[22] Looking back over this time, Lieutenant Governor William Bross, in office from 1865 to 1869, recalled that the water supply was often muddy, and that many families, given the limited scope of the public supply and the problems associated with private wells, were forced to get their water by the bucket or the barrel from the water carts.[23]

During the fall of 1847 the quality of the water supply was particularly objectionable. The following spring the Chicago Mechanics' Institute appointed a committee, consisting of S. D. Childs, A. F. Bradley, and W. H. Kennicott, to devise a better plan for getting a purer supply. In May the committee proposed that a pipe be laid 3 feet below the lake's low-water mark, and that it extend from a point opposite First Street to a point where the lake was 20 feet deep. The idea was to attach an elbow to the pipe's eastern end and have the pipe extend upward 10 feet from the bottom to draw water that would be equally free of floating impurities and of residue from lake-bottom disturbances. The intake pipe would continue down the center of First Street and extend from there to the western boundary of the city. At every other street crossing, wooden pipes would be laid to distribute water throughout the service area. It was estimated that the wooden pipes would cost about $2,000 per mile.[24]

The suggested plan included two reservoirs, both to be erected on the lakefront, and each capable of holding 20,000 cubic feet of water. The plan also suggested a better way of ensuring purer water by first having it brought into the reservoir nearest the lake, where the water would be drawn off from the top and carried by an elbow to the bottom of the second reservoir. From there, the water would be drawn off again near the top of the second reservoir and then distributed through the wooden pipes. This plan contrasted with the system in use at the time, which drew water from the bottom of the lake, poured it into the top of the reservoir, and then took it out from the bottom of the reservoir, where sediment naturally had settled. This plan was to form the template for later efforts.[25]

* * *

In 1832, ships from the east had unleashed Chicago's first cholera outbreak. Now, nearly twenty years later, and despite nascent efforts to procure clean water, the city endured its second great epidemic, which arrived this time from the south, along the very waterway that was to contribute so prominently to the city's early growth. On April 29, 1849, the canal boat *John Drew*, carrying emigrant workers, arrived from New Orleans, having traveled up the Illinois and Michigan Canal. Its captain, John Pendleton, died of cholera within hours of the boat's arrival. Many others on board were sick, too, but in the absence of quarantine regulations for arriving vessels, they were allowed to disembark, and the disease quickly began to spread.

The filth theory still guided public understanding of the etiology and transmission of disease, and so the city reacted as best it knew how. One hundred barrels of lime were scattered on the streets, vacant lots were cleaned, slaughterhouses were prohibited within the city limits, and the construction of the first sewers was authorized. On May 15 an appropriation was made for the first cholera hospital to house and treat the many poor residents who were afflicted by the disease.[26]

Despite these efforts, the disease was relentless. Between July 25 and August 28, there were 1,000 reported cases of cholera and 314 deaths. The cholera hospital closed in September, only to reopen in October under a new wave of cases. By the end of the 1849 outbreak, 678 people had died of the disease, an incredible rate of 1 in 36 members of the population. Nor was this second deadly epidemic of cholera over. The disease reappeared in 1850, and within a three-month period, from July to September, it killed another 420 people.[27]

Although the Chicago Hydraulic Company had made a significant step forward in supplying the young city with water, recurring issues of poor water quality and widespread fear over cholera led to a civic outcry for a better solution. There were also persistent complaints about the limited scope of the water distribution network. By 1850, the Chicago Hydraulic Company still supplied only part of the south side of the city and an even smaller portion of the west side. The rest of the city was obliged to rely on private wells and water carts, and it was in parts of the city where private wells were most prevalent, especially the north side, that the cholera outbreaks and their accompanying mortality rates were the worst. Even the water supplied by the watermen, who drew from the lake and the river, grew increasingly suspect, given that their water was taken from, in the words of the *Chicago Tribune*, "the common receptacle for all the filth of the city."[28]

In 1850 the Chicago Hydraulic Company laid 1 mile of pipe, mostly on the west side, bringing its total distribution system to 9.5 half miles, with a total of one thousand taps. Of this number, approximately eight hundred were used by families, with the rest used by stores, public houses, livery stables, and other commercial concerns. Thus, at its high mark, the Chicago Hydraulic Company was serving only about one-fifth of the population, which by

1850 had reached nearly 30,000.[29] It was time for a change, and both the public and the press began to argue more forcefully that the welfare and prosperity of the city demanded a "cheap and abundant supply of pure water to the entire population."[30]

In the beginning, individuals had relied on themselves to provide their own water. Then, as was the Chicago way, entrepreneurs stepped in to try to fill the need. Each approach had its time and place. But now, given the limitations of distribution, the poor water quality, and the horrifying persistence of cholera, residents demanded that the city itself provide a comprehensive response. This demand reflected a deeper shift in the consciousness of the city's residents, who had begun to think of themselves less as a collection of isolated individuals and more as a *polis* encompassing shared needs and aspirations.[31] The city was coming into its own, and as part of this maturation it was the city's turn to lead the quest for pure water.

Chapter 2

✦

The City Takes Over

In April 1850, a town hall was held to discuss the future of the city's water supply. A committee was formed, chaired by Peter Page and comprising representatives from the city's various areas: R. H. Foss and T. M. Moody from the south side, Alson S. Sherman and Luther Marsh from the west side, and R. J. Hamilton and William E. Jones from the north side. After research and deliberations, the committee recommended that the city take over responsibility for supplying its citizens with water through a city-owned and -operated system. Early the following year, on February 15, 1851, the state legislature—at last formally recognizing the city's authority over its own infrastructure—authorized the creation of a board of water commissioners for the city and incorporated the Chicago City Hydraulic Company.[1]

The first board of water commissioners was made up of John B. Turner, Horatio G. Loomis, and Alson S. Sherman. They began their work on July 16, 1851, and ten days later hired William McAlpine to oversee the design and construction of the new waterworks. McAlpine had come to Chicago from an already distinguished background in his native New York, where he had broad experience as an engineer. He had been responsible for designing and building the Albany waterworks and in later life would be involved with the design and construction of the Brooklyn Bridge and the George Washington Bridge. He served as president of the American Society of Civil Engineers from 1868 to 1869 and was the first American elected to membership in Great Britain's Institution of Civil Engineers.

The board's instructions to McAlpine were perhaps enough to have dissuaded a less competent man. He was instructed to make examinations of the lake in the vicinity of the city so as to determine the best possible way of providing water for the populace. He was to "regard [the] economy and permanency" of the proposed supply, and, to aid him in his deliberations, the board provided its own projection that the city's population in 1866 would be "one hundred thousand souls."[2] At the time this prediction was made, in 1851, the population was about 35,000. The commissioners likely thought they were being expansive in their estimate of Chicago's growth, but by 1870 the city's population would be just shy of 300,000.

In making its charge to McAlpine, the board emphasized the benefits of an abundant supply of pure water, focusing especially on its health benefits. It also noted the need for a dependable water supply for the fighting of fires and noted that should "a plan of sewerage be adopted by the City, the water will be an active agent . . . for ridding the City of the impurities which sewers are intended to remove."[3] In these matters the board was prescient, but little could its members imagine the roles that disease, sewerage, and fire would play in Chicago's history in the decades to come.

McAlpine embraced his assignment with focus and verve. He received the board's charge on July 14, 1851, and on July 23 he submitted a preliminary memorandum to further define the project. He then returned to New York while his able assistant, E. W. Smith, remained in Chicago to take soundings in the lake, where various possible inlets had been suggested, as well as at possible locations for the laying of water pipes that would extend to the various divisions of the city. Smith also provided McAlpine "with very beautiful designs for the engine house, and the buildings for supporting the reservoirs, as well as aided in the arrangement and *computations* of the other plans for the work."[4]

Popular sentiment opposed the drawing of any lake water south of the Chicago River's mouth because of the belief that a southward river current brought pollution from the river to the lake's southern shoreline. But McAlpine conducted his own evaluation and concluded that it was "very questionable whether the small quantity of water which is discharged from the river, would affect the quality of water in the lake, at a point 1.5 miles south."[5]

Using his own observations and the data and advice provided by Smith, McAlpine devised a general plan whereby water would be drawn from the lake to a pump well, from which it would then be pumped by steam engine through iron mains to centrally located reservoirs and then distributed throughout the city by way of iron pipes. He considered four different locations for the pumps and reservoirs:

(1) Locate engine and pumps on the lake shore north of river near the end of Chicago Avenue with water forced into a single reservoir also situated on the North Side;
(2) Locate engine and pumps on the lake south of the river near 12th Street with water forced into a single reservoir on the South Side;
(3) Locate engine and pumps as in the first plan, but with water forced into three reservoirs servicing the North, South, and West Sides; or
(4) Locate engine and pumps as in the second plan with reservoirs placed as in the third plan.[6]

In addition, McAlpine proposed that a wooden crib be placed 600 feet from shore to serve as the inlet to a wrought-iron pipe 30 inches in diameter, which would carry the water to the pump well. The pipe would be placed in a trench, to be dug at least 3 feet below the lake's floor, which would nowhere lie less than 5 feet below the water's surface. The pipe would be covered with stone, and at the pump well it would turn down to the bottom of the well.[7] McAlpine completed his proposal to the Board of Water Commissioners with a description of the engines, pumps, reservoirs, and distribution pipes that would accompany each of the four plans.

After consideration, the board adopted a modified version of the third plan, by which water would be taken from the lake near the termination of Chicago Avenue and pumped to three reservoirs, one for each division of the city. The plan called for a 600-foot intake pipe leading to a well and engine house constructed on or near the beach. From there the water would be pumped by a 170-horsepower engine into the reservoirs, to a height 85 feet above the surface of the lake (about 75 feet above the general level of the city). The design called for iron to be used for the distribution pipes as well as for the tanks for the reservoirs.[8]

The plan, as adopted by the board, would provide about 48 miles of distribution pipe and a pumping capacity of 3 million gallons in twelve hours. It was estimated that this approach would provide for the needs of a city of 100,000 residents. The estimated cost of the plan was $570,000, which some considered extravagant, given what they thought was an inflated expectation of the city's population growth. Nevertheless, when the plan went to a vote at the municipal election of March 2, 1852, only 513 of 4,445 votes were cast against the plan (1,244 voters apparently had no opinion one way or the other). Thus the city, on its own authority, adopted its first comprehensive waterworks plan.[9]

Site selection and engineering were not the only issues facing the new waterworks. The old Chicago Hydraulic Company objected to losing its seventy-year charter and threatened legal action to halt the new plan. Eventually the old company softened its objections and offered to settle if the municipal waterworks would buy out its fixtures and franchise rights. After considerable sparring, the two parties reached an amicable agreement, and construction of the new waterworks began in the summer of 1852.

The year 1852 saw success as well as failure. The pump well was built, along with the foundations for the waterworks building and tower, and a portion of the inlet pipe was laid. But the temperamental lake proved too rough for McAlpine's proposed 600-foot pipe and wooden crib. After several failed attempts, the idea was scrapped and replaced with a plan for an inlet pipe, much closer to shore, in an inlet basin protected by a semicircular breakwater with an opening to the southeast.[10] This compromise solution, which brought the inlet much closer to shore and only 3,000 feet from the mouth of the river, would prove problematic in years to come.[11] In 1852, however, the first iron distribution pipes, imported from Scotland, were laid, replacing the wooden pipes that had been used before. Also in 1852, a report

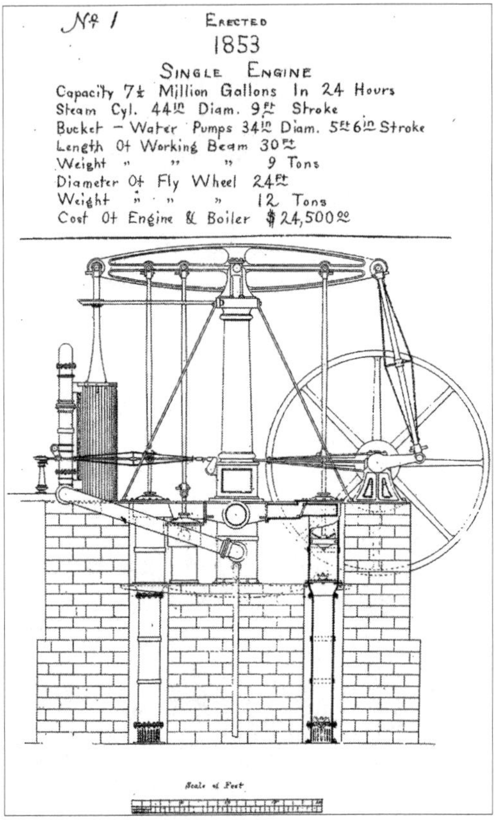

Figure 3. Engine 53. Source: City of Chicago, *First Annual Report of the Department of Public Works to the City Council of the City of Chicago for the Fiscal Year Ending December 31, 1876* (Chicago: Clark & Edwards, 1877).

on cholera by City Physician A. B. Palmer identified the highest incidence of the disease in areas that still relied on private wells. On the basis of his findings, Dr. Palmer advocated a lake-based water supply and better drainage.[12]

In 1853, about 80 percent of the city's population still got its water from private wells and water carts.[13] Progress on the waterworks continued throughout 1853, and in the fall of that year the standpipe was erected and the engines were installed. The buildings and installations were quite substantial. The pump well was rectangular, 20 by 30 feet, and 25 feet deep from the floor of the engine house to the bottom of the well. The walls were stone, 6 to 7 feet thick, and the engines were set on those walls. The buildings themselves were brick, 40 by 50 feet, and comprised two wings for the boiler rooms, each of which was 30.5 by 40.5 feet. The water tower was also brick, 136 feet tall, 14 feet square at the base, and 11 feet square at the top. The

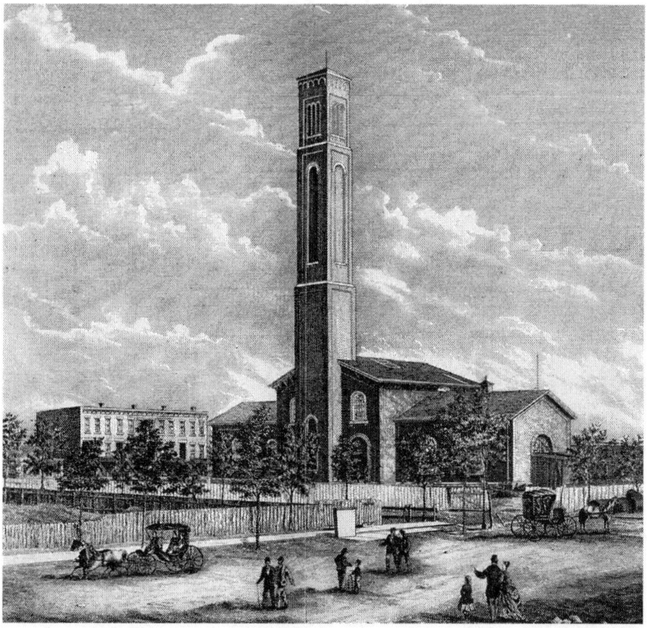

Figure 4. Chicago Avenue pumping station in 1853. Source: City of Chicago, *Eighth Annual Report of the Board of Public Works to the Common Council of the City of Chicago for the Municipal Fiscal Year Ending March 31st, 1869* (Chicago: Jameson and Morse, 1869).

interior of the tower was divided by a wall with a smoke chimney on one side and the iron standpipe on the other.

The foundation rested on a bed of sand, and at one point during construction the tower leaned 14 inches off vertical. It was eventually made plumb, however, and remained so until its demolition. The original pumping works consisted of a vertical-beam condensing engine on the north side of the building, with a steam cylinder 44 inches in diameter and a stroke of 9 feet, which powered two single-action pumps, each 34 inches in diameter and with a 5.5-foot stroke. On the south side of the building was a horizontal-beam noncondensing engine with a steam cylinder 18 inches in diameter and a 6-foot stroke, which powered one double-action pump of the same dimensions.[14] (A noncondensing steam engine discharges steam from the cylinder at a pressure nearly equal to or above atmospheric pressure; in a condensing steam engine, the steam is condensed to below atmospheric pressure so as to gain the maximum amount of energy.) The condensing engine, installed in 1853, was referred to both as Old Sally and simply as 53. It had a pumping capacity of 7.5 million gallons every twenty-four hours. It was started for the first time on December 16, 1853, to test the pipes, with the water running through the first-laid 4-inch iron distribution pipe to an open fire hydrant on North Clark Street.

Overseeing the new pumping station was twenty-four-year-old DeWitt C. Cregier. At that youthful age, Cregier likely never would have guessed at the long and honorable service he was to provide for a city that would come of age in his own lifetime. He remained in charge of the Chicago Avenue pumping station until 1879 and then served as city engineer from 1879 to 1882 and as commissioner of public works from 1882 to 1886. He capped his remarkable career in public service as mayor of Chicago from 1889 to 1891. A gifted engineer, he designed several pumps and held patents to the so-called Cregier hydrant, which incorporated a fire hydrant, a drinking fountain, and an animal-watering basin.

During the earliest operation of the waterworks, there were no available reservoirs, and so water was allowed to run through open fire hydrants to keep 53 pumping. The first permits to take water from the waterworks were granted on February 12, 1854, and water was introduced into buildings for the first time on February 15, 1854. For the first four months, water was supplied only nine hours per day, with no water on Sunday except for firefighting. After the first four months, the supply was available twenty-four hours per day.

Three reservoirs were constructed during 1853 and 1854: one at LaSalle Street and Adams Street (1853), one at Chicago Avenue and Sedgwick Street (1853), and one at Morgan Street and Monroe Street (1854). Each was capable of holding two or three days' supply. By the end of 1854, 30.5 miles of distribution pipe had been laid.[15]

The first years of the waterworks were not without difficulty. The inlet pipe, laid too close to shore, was exposed to wave action, and on one occasion it was completely clogged because its strainer was full of insects. In early 1854, a river pipe at State Street was broken by an anchor dragging from a vessel. On November 22, 1854, the Adams Street reservoir, which had been filled to a depth of 28 feet, was discovered to have cracks in its masonry on all sides from the pressure of the water. Temporary repairs were made, but it was not until June 1855 that the reservoir was sufficiently reinforced to be able to hold 18 feet of water. In December of the same year, a large crack in the main pipe near the standpipe of the waterworks occasioned more repairs.

The quality of the water supply, given its source so close to shore, was not good. During storms the water would be muddy, but, even worse, in spring and early summer the water supply was plagued by small fish that ended up in washbowls or got stuck in faucets. One early resident noted that when the fish found their way into a hot water tank they would get "stewed up into a very nauseous fish chowder."[16] The "fish nuisance" continued for years, and in April 1855 the *Chicago Tribune* complained that unless a better strainer were installed in the waterworks, there soon would be no fish left in the lake, and the paper warned readers to be "careful drinking not to get fish bones in your throat."[17] A contemporary history went even farther, proclaiming that the city's water was flavored with fish and that unscrupulous bar owners had

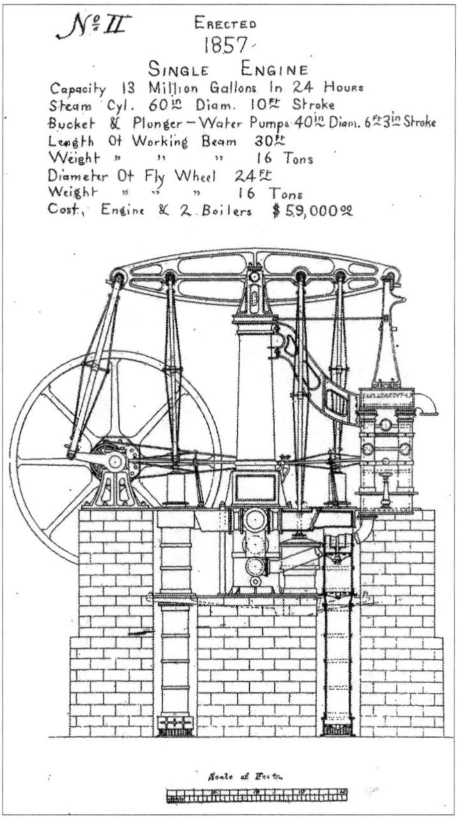

Figure 5. Engine 57. Source: City of Chicago, *First Annual Report of the Department of Public Works to the City Council of the City of Chicago for the Fiscal Year Ending December 31, 1876* (Chicago: Clark & Edwards, 1877).

been caught in the act of adulterating their liquor when "pickled minnows" were found in their offerings.[18] To help combat the muddiness as well as the fish and debris finding their way into the system, a breakwater was built, and the inlet area was dredged in 1855. By the end of that year, $496,849.64 had been spent on construction and repairs to the waterworks. Over 41 miles of pipe had been laid, and 4,251 buildings were being supplied with water.[19]

By early 1856, demand on the engines of the waterworks had reached maximum capacity. In 1857 a new engine, of similar design to the one called 53, was installed to replace the noncondensing engine. It was not an easy installation, in part because of the difficulty of building a new foundation while the existing engine was in use, and in part because the position of the tower did not allow an engine of normal design. The new engine, called 57, had to be built with its valve gear on the side instead of in the front so as to allow installation in the existing space. The new engine began operating in July 1857 with a much increased capacity of 13 million gallons every twenty-four hours, as compared to the old engine's 7.5 million daily gallons. By mid-1857 the waterworks was supplying 7,053 buildings.[20]

The city established a rate structure for the new water supply and based it on anticipated demand. As had been the case with the earlier private waterworks company, private homes paid the least, with larger amounts paid by hotels, boardinghouses, and manufacturing companies. It became a source of great pride for the city that the waterworks would always be financially self-sustaining, paid for by water receipts and bonds. Table 1 summarizes the finances for the early years of the city's waterworks.[21]

Table 1. Financial summary for the waterworks, 1854–61

YEAR	COST OF WORKS ($)	OPERATIONAL EXPENSE AND INTEREST ($)	REVENUE ($)
1854	393,045.32	38,128.51	26,808.56
1855	496,849.64	59,051.27	54,739.19
1856	641,509.93	73,087.23	76,806.36
1857	738,436.51	85,170.61	97,008.55
1858	912,152.23*	94,730.09	102,709.32
1859	989,595.23	108,809.48	123,600.77
1860	1,013,245.06	111,718.99	131,536.34
1861	1,074,338.94	118,414.43	131,035.10
TOTALS	3,077,179.23	689,110.61	744,244.19

*Includes $82,974.28 for building of north and west reservoirs.

The distribution system and pumping capacity of the waterworks increased steadily from 1857 to 1861. By January 1861, a total of 91 miles and 127 feet of pipe had been laid. In 1860, a daily average volume of about 4.7 million gallons of water was being pumped, representing an increase of 21 percent over the volume pumped in 1859. In turn, the volume of water pumped in 1859 represented an increase of 29.5 percent over the volume pumped in 1858. Despite the city's best efforts, however, it was obvious that capacity was not keeping pace with the city's explosive growth.

Problems also persisted with the quality of the water supply, as we will see more fully in chapter 3. Muddiness, debris, and fish continued to make the water unsavory, if not necessarily unsafe. In 1859, R. J. Hamilton of the Board of Water Commissioners first broached the idea of extending the inlet pipe farther out into the lake, to a distance of 1 mile. In 1861, however, the board submitted its summary of an informal analysis that had been conducted in 1860. In its summary, the board reported that no discernible difference had been found between lake water taken 1 mile from shore and water taken from an onshore hydrant.[22] Nevertheless, the board decided to seek another opinion.

In March 1860, the Board of Water Commissioners asked the chief engineer of the Board of Sewerage Commissioners to review the idea of extending the inlet pipe. During his investigation, the engineer determined that Chicago's average daily water consumption for 1860 had been about 5 million gallons, with demand on some days in July reaching 7 million gallons. To estimate future needs, he doubled those volumes and calculated that a 5-foot-diameter iron pipe 1 mile long could indeed deliver the desired 10 to 14 million gallons every twenty-four hours.[23] Given the cost of extending the pipe into the often tumultuous lake, however, not to mention the practical considerations of such an effort, the engineer found this idea untenable. And yet the proposal he offered instead, in a single sentence, eventually led to one of the great engineering feats of the nineteenth century: "To avoid the various risks and uncertainties attendant upon the laying and maintaining of the proposed pipe, it has been thought a *tunnel* might be made."[24] The engineer's name was Ellis Sylvester Chesbrough, and the emphasis on *tunnel* was his.

Chapter 3

✦

The Engineer

Despite the efforts of the Chicago City Hydraulic Company, cholera persisted in the city. There had been a four-month outbreak in 1852, but when 1853 passed without further incidence of the disease, the city cautiously hoped that the scourge had been eliminated. Then came 1854, and, in the words of the *Chicago Tribune*, "the demon reappeared."[1] The first cholera cases of that year appeared in April but were either explained away or simply denied by the press, the public, and even the Board of Health. But in June and July the worst of the outbreak hit, and people began to die at the rate of 60 per day. The *Chicago Tribune* reported that "the death cart was seen continually in the streets . . . [and] the streets seemed full of hearses and coffins."[2] Chicago was in the grip of yet another cholera epidemic, and this time the toll would be almost unimaginable. At first the deaths came so rapidly that an accurate record of them could not be made. By the end of the year, the disease had killed 1,424 people.[3]

Despite the devastating impact of the 1854 cholera outbreak, at least one newspaper of the time tried to continue downplaying the deadly reality. In June 1854 the *Chicago Tribune* attempted to blame the death toll on "the foreign population" of the city, concluding that, other than the "usual summer complaints, we can insure the public that our city is enjoying good health."[4] Such misplaced boosterism notwithstanding, it was clear that it was time for the city to address the main cause of its repeated cholera outbreaks—the lack of a workable sewerage system, a deficiency that in turn led to the vicious cycle of impure drinking water. In response to a public outcry, Mayor Isaac Milliken called a public meeting on January 12, 1855, so that residents could hear the report of the Common Council's Committee on Sewerage. At that meeting, an act to incorporate a board of sewerage commissioners for the City of Chicago was presented, and on February 14, 1855, the state legislature formally created the Chicago Board of Sewerage Commissioners. The newly formed board was instructed to "examine and consider all matters relative to the thorough systematic and effectual drainage of the city of Chicago . . . as best to promote the healthfulness of said city."[5]

In May 1855 the Common Council appointed William B. Ogden, Joseph D. Webster, and Sylvester Lind as the city's first sewerage commissioners.

Figure 6. Ellis Sylvester Chesbrough. Source: *Biographical Sketches of the Leading Men of Chicago, Photographically Illustrated by J. Carbutt* (Chicago: Wilson & St. Clair, 1868).

They in turn set out to find "the most competent engineer of the time who was available for the position of chief engineer."[6] Their selection was Ellis Sylvester Chesbrough, who at the time was serving as city engineer in Boston. At the invitation of Ogden, then president of the Board of Sewerage Commissioners, Chesbrough accepted the position of chief engineer on October 1, 1855. He resigned his position in Boston and came to Chicago in December. Upon his arrival, he submitted to the sewerage commissioners a pioneering report that contained a plan for the first comprehensive sewerage system for a major city in the United States.[7] It eventually revolutionized how sewerage was thought about and managed, and it established Chesbrough as the foremost sanitary engineer of his time.

Chesbrough began with the premise that "the main object of the sewers is to improve and preserve the health of the city."[8] The old system of privy vaults had to be eliminated, said Chesbrough, and replaced with a sewer system that would accept both surface water and sewage. He enumerated four alternatives for ridding Chicago of its sewage. The sewage could be (1) dumped into the Chicago River and thus into Lake Michigan, (2) dumped directly into Lake Michigan, (3) stored in reservoirs and then pumped out for use as fertilizer, or (4) dumped into the Chicago River and then transferred by way of a canal to the Des Plaines and Illinois Rivers.[9] Although he believed that a canal was the optimal solution—the first of many prescient ideas to come from this gifted engineer—he opted for the quicker and more economical option of dumping directly into the river and the lake. He understood that this alternative would be problematic for the city's water supply, but his hope was that the sewage would be sufficiently diluted by the river and the lake to eliminate its threat to the supply of drinking water. Unfortunately, however, that hope was never realized, and eventually Chesbrough had to find a way to supply the city with pure water.

Because there were no US cities on which to model a plan, in December 1856 the Board of Sewerage Commissioners sent Chesbrough to Europe to examine how sewerage was handled there. Upon his return, on March 25, 1858, Chesbrough issued a report in which he reviewed his analysis of sewerage management in Liverpool, Manchester, Rugby, London, Amsterdam, Hamburg, Berlin, Paris, Worthing, Croydon, Leicester, Edinburgh, Glasgow, and Carlisle, and he concluded that the optimal solution for Chicago at that time was a system of combined, intercepting sewers that would discharge into the Chicago River.[10] At the end of his report, however, Chesbrough returned to his earlier idea that a "thorough cut for a steamboat canal, to the Illinois river, which the demands of commerce are calling more and more loudly for, if ever constructed, would give . . . perfect relief to Chicago."[11] This deepening of the existing Illinois and Michigan Canal would reverse the flow of the Chicago River, carrying sewage westward, away from the city and its water supply. In 1871, as we will see, Chesbrough got to try his hand at this solution, but not until the early twentieth century would the final reversal of the Chicago River become a reality.

Chesbrough's 1858 report was the first thorough and exhaustive study devoted to a systematic sewerage system by an American engineer. Benezette Williams, who succeeded Chesbrough as Chicago's city engineer, later remarked, "It is hard for us to appreciate the importance of this report and the effect it exerted not only upon the destiny of Chicago, but also upon other cities of the country."[12] By adopting Chesbrough's approach, Chicago became the first American city to have a systematic sewerage system, and Chesbrough became the nation's recognized authority on sanitary engineering.

Over the decade that followed his 1858 report, Chesbrough's sewerage plan became a reality. It called for the raising of the city—a dramatic public works project encompassing the elevation of stores and hotels that remained in operation throughout the process—in order to create a grade sufficient for the functioning of gravity-fed sewers.[13] But even that astonishing feat would pale in comparison to Chesbrough's unprecedented plan for securing a source of pure drinking water for the city.

* * *

Ellis Sylvester Chesbrough was born in Baltimore County, Maryland, on July 6, 1813, the first son of Isaac M. Chesbrough and Pharina Jones.[14] His first paternal ancestor in America was William Chesebrough, who came to the Massachusetts Bay Colony in 1630 and would later become cofounder of the town of Stonington, Connecticut. Isaac Chesbrough, a farmer, tried his hand at other business ventures soon after Ellis's birth. Unfortunately, Isaac was not successful in business, and young Ellis was forced at the age of nine to forgo the liberal education his father had hoped to provide for him. Instead, Ellis had to go to work and help support the family. Between the ages of nine and fifteen he received only about a year of formal schooling. The balance of his time was spent at work, which included service in the mercantile houses of Baltimore.

In May 1828, when he was almost fifteen, Ellis took his first step toward fulfilling the legacy that fate held in store for him when he became a chainman with a party engaged in a preliminary survey for the Baltimore & Ohio Railroad Company. He quickly showed his abilities. The officers in charge were impressed by his diligence and by his practice of taking notes as the work progressed. By his second year on the job, he was occasionally put in charge of his own crew.

It was a propitious beginning for young Chesbrough. Colonel Stephen H. Long was at that time chief engineer for the Baltimore & Ohio Railroad, and Chesbrough's party was under the command of Lieutenant Joshua Barney. Many of the engineers on the road were army officers and graduates of the US Military Academy's engineering program, which emphasized practical as opposed to theoretical training. They welcomed their young apprentice and afforded him every opportunity to acquire knowledge about the principles and science of engineering.

The road to his future success continued in 1830, when Chesbrough left the employ of the Baltimore & Ohio Railroad to work as assistant engineer for the State of Pennsylvania on the survey for the Allegheny Portage Railroad. Once again Colonel Long was chief engineer. In 1831, Chesbrough served as assistant engineer under Captain William Gibbs McNeil for the Patterson & Hudson River Railroad, and then he worked on the location and construction of the Boston and Providence main line and on the Taunton & New Bedford branch line. In 1836, he was resident engineer under the supervision of Lieutenant George W. Whistler on a proposed railroad line from Lowell, Massachusetts, to Concord, New Hampshire. For the next several years, until 1842, Chesbrough was employed on surveys and construction of the Louisville, Cincinnati & Charleston Railroad, part of the time as senior assistant to Captain McNeil and then, after McNeil resigned, as acting chief engineer. During that time, the rail line was built from Charleston to Columbia, South Carolina. Among the engineers under Chesbrough's charge on that road were George S. Greene (a future Civil War general as well as a future president of the American Society of Civil Engineers), J. P. Kirkwood and Alfred W. Craven (who would go on to become members of the American Society of Civil Engineers), and John C. Fremont (like Greene, a future Civil War general).

The financial crisis of 1837, and the resulting depression, which lasted into the 1840s, severely restricted funding for the country's internal improvements, and many engineers were thrown out of work. Between 1837 and 1844, if one takes into account the New York & Erie Railroad, Indiana public works, the Illinois and Michigan Canal, the Chesapeake and Ohio Canal, the Baltimore & Ohio Railroad, and Ohio public works, the total number of employed engineers dropped from 116 to 13.[15] Chesbrough, caught up in this widespread reduction, lost his employment in 1842.

Chesbrough had married Elizabeth A. Freyer on December 23, 1837. Their first child, Isaac Sumpter Chesbrough, was born on January 19, 1839, and their second, Henry Freyer Chesbrough, was born on September 19, 1840, in Totness, South Carolina (Henry's own son, Harry, would follow in his grandfather's footsteps and also become a civil engineer). In 1842, Chesbrough moved with his family to Providence, Rhode Island, where his father then lived. He spent the following winter in the workshops of the Stonington Railroad Company, learning the practical use of tools. His third son, Lewis Collins Chesbrough, was born on September 5, 1842 (Lewis died on February 29, 1844). In 1843, in Niagara County, New York, Chesbrough purchased a farm and made what proved to be an unsuccessful attempt at farming before returning to engineering in 1844. Between 1844 and 1846 he was engaged successively in the construction of the Stoughton branch of the Boston & Providence Railroad, the Ashburnham and Brattleboro line, and the Pawtucket branch of the Boston & Providence Railroad.

And so between 1828, when he was fifteen, and 1846, when he was thirty-three, Chesbrough acquired a solid foundation as an engineer through extensive on-the-job training with the railroads. His training spanned a transformative period in the history of civil engineering in America, when there were two major training sources for engineers—the US Military Academy at West Point, and the New York State canal system, sometimes referred to as the Erie School of Engineering.[16] Both the military academy and the canal system stressed a practical, hands-on approach that suited Chesbrough. His industry, diligence, and natural abilities led him to a succession of mentors, who schooled him in all aspects of engineering as it pertained to the construction of the expanding railway network, including the work of grading, tunneling, and surveying.

As a further indication of his character and skill, in 1846, with nothing but railroad training behind him, Chesbrough was offered the position of chief engineer for the Boston Water Works, with responsibility for overseeing construction of the Cochituate Aqueduct. The project was divided into two divisions, western (running 14.6 miles from Cochituate Lake to the Brookline reservoir) and eastern (eventually including the pipes from the reservoir to the city). Chesbrough was to be in charge of the western division at a salary of $3,000 per year, with William S. Whitewell employed to oversee the eastern division.[17] Because Chesbrough was largely unacquainted with hydraulic engineering, he was reluctant at first to accept a position so far outside the scope of his prior training. At the prodding of his friends, however, he finally agreed to take the job, but only after being assured that John B. Jervis, who had planned the aqueduct, would remain as a consulting engineer. Chesbrough oversaw the construction of the aqueduct until its completion in 1849. In 1850 he was made sole commissioner of the waterworks, and in 1851 he was appointed the first city engineer for Boston.

Chesbrough found in Jervis a mentor and kindred spirit who would help lead him to his place in the pantheon of civil engineering. Although the two men were widely separated by age—Jervis was eighteen years older—they followed remarkably similar and parallel paths as engineers. Chesbrough had begun his training with the railroads under the tutelage of engineers in the military tradition, whereas Jervis had begun under the tutelage of engineers from the New York State canal system. Jervis had been born in Huntington, Long Island, but he grew up, as fate would have it, in Rome, New York (at the time called Fort Stanwix), the site of a navigable canal connecting the Mohawk River with Wood Creek.[18] As a boy, Jervis was educated in the common schools of the day, and he worked in his father's sawmill.

In 1817, a family friend, Benjamin Wright, chief engineer for the Erie Canal and in need of an axman, approached Jervis's father. Young Jervis got the job, beginning at the bottom tier in the canal system just as Ellis Chesbrough would later begin as a chainman for the railroads. Like Chesbrough, Jervis progressed quickly, and within two years he was made resident engineer on a 17-mile stretch of the canal. He worked on the Erie Canal for the next seven years, gaining a solid foundation from some of the leading engineers of the day.

In 1825, following his mentor Benjamin Wright, Jervis became chief engineer of the Delaware and Hudson Canal Company. During his tenure there, Jervis brought the first locomotive to America from England. The canal and the accompanying railroad were completed in the fall of 1829, and during their initial use Jervis employed John H. McAlpine to superintend the construction of machinery. John McAlpine introduced his son, William, to Jervis. William McAlpine studied under Jervis until 1836 and, as we have seen, was hired in 1851 by the newly created Chicago Board of Water commissioners to oversee the design and construction of the new Chicago waterworks. Jervis had already become a thread in the tapestry of Chicago's quest for pure water.

In 1830 Jervis, reversing Chesbrough's route from the railroads to water, became chief engineer of the Albany & Schenectady Railroad, the first railroad constructed by the State of New York. He subsequently became chief engineer of the Schenectady & Saratoga Railroad, and while serving in that capacity he designed a new railway engine that would become the standard design for American locomotives.

After completing his work on the two railroads, Jervis returned to water, working as chief engineer on the Chenango Canal in 1833 and on the widening of the Erie Canal in 1835. In 1836 he became chief engineer of the Croton Aqueduct, which would become one of the nineteenth century's greatest examples of hydraulic engineering. He remained at work on the Croton Aqueduct until its completion, in 1848. He took time from those duties, however, to accept an appointment in 1846 as consulting engineer for the Cochituate Aqueduct in Boston. It was there that Jervis met Chesbrough, and the two men remained friends and colleagues for the rest of Jervis's long and exceptional life.

And so Ellis Chesbrough became a beneficiary of the two main engineering traditions of the nineteenth century. From the military tradition and his early railroad mentoring by Stephen H. Long and others, Chesbrough learned the skills of surveying, tunneling, and grading. (Indeed, it is even possible that Chesbrough acquired from Long some knowledge of the topography of the Chicago region because in 1816 Long had been tasked with exploring the area between the Illinois River and Lake Michigan. Long's explorations included such tributaries of the Illinois River as the Des Plaines, Du Page, and Kankakee Rivers. He saw early on the potential for a connection between the Des Plaines and Chicago Rivers and recommended, in the report of his explorations, the building of a canal between the two.)[19] From the New York canal system (the so-called Erie School of Engineering), and especially from his work with John B. Jervis, Chesbrough learned hydrology and the intricacies of water supply.

* * *

When Chesbrough arrived in Chicago, in 1855, the city was at a crossroads. On the one hand, Chicago was already established as the major transportation hub of America. Its unique topography at the edge of the subcontinental divide made it a natural bridge for the resources and products of the West and the markets of the East. But, on the other hand, the city's exponential growth also threatened to be its undoing. For the city to endure, it had to find an answer to its defining dilemma: how to deal with its waste while also finding a stable supply of clean water. What had once been a matter of convenience was becoming a matter of survival.

Chesbrough's experience had singularly prepared him for the challenges facing Chicago. The engineering involved in meeting those challenges would be wide and varied, but they would require more than just technical skill. Chesbrough would prove to be an engineer of vision, ingenuity, and imagination, perceiving solutions overlooked by more timid minds. He possessed an uncanny ability to see the basic issues at the heart of big problems, and to provide solutions of deceptive simplicity. And once an answer presented itself to him, he was dogged in its pursuit. He matched vision with perseverance, and his answers to Chicago's dilemma of waste and water would set an example of engineering excellence for an entire nation.

Chapter 4

✦

The Chesbrough Plan

Between 1856 and 1860, Chesbrough's sewerage plan progressed, and the city was slowly raised to accommodate the laying of the new sewer lines. The good news was that the sewerage system worked as planned and successfully drained the city's waste with increasing efficiency into the Chicago River and then into the lake beyond. The bad news flowed directly from the good. By 1860, the vicious cycle of trying to use the same sources of water to dispose of the city's waste and supply its drinking water had come to a critical juncture. And it was at this intersection of competing interests that Ellis Chesbrough found himself.

Chesbrough reported in April 1861 that the sewers were having a positive effect on the lingering persistence of cholera. With the increased use of sewers, the problem of contamination from privy vaults seeping into private wells was decreasing. Chesbrough provided statistics showing a marked decline in mortality due to cholera. In 1846, the death rate in the city was 23 per 1,000 cases. That number increased to 53 per 1,000 in 1849, decreased over the next few years, and then spiked again to almost 54 per 1,000 in 1854. From that time forward, with the construction of the sewerage system, there was a steady decline in mortality until 1860, when the death rate was just shy of 19 per 1,000. That rate would remain fairly constant until the early twentieth century.[1]

When Chesbrough, in his 1855 report, stated that the most expedient solution to Chicago's problem of waste disposal would be to drain the city's waste and sewage directly into the Chicago River, he had fully recognized the drawbacks. The most obvious was the potential health hazard, but, given the river's slow and sometimes almost nonexistent current, there were also concerns that sediment deposits could eventually obstruct navigation in the river. Chesbrough addressed the health concerns by proposing feeder canals that would flush the north and south branches of the river. He rejected the concerns about the potential for sediment to obstruct navigation as "groundless" and concluded that any sediment that might be deposited would be flushed away during periods of rain and flood.[2]

By 1858, however, after his exploratory trip to Europe, Chesbrough had reconsidered his proposal to flush the river by way of canals from the lake to the north and south branches. His experiences in Worthing, England, and along the coast near Liverpool, gave him pause. In those locations, sewage had been discharged into the sea, but Chesbrough had learned from local reports that "the mere discharge of filth into the sea gives no security against its being cast back in a more offensive state than ever."[3] He worried that if sewage were similarly discharged into the lake, it might simply be carried south along the lakeshore, thus "creating on the lake shore as great a nuisance as would be taken from the river."[4] Accordingly, he recommended that the idea of flushing the river with feeder canals be put on hold. Instead, he suggested that the pumping station at Bridgeport could pump additional water to accomplish a similar purpose. This approach would also provide data by which to judge the effects of the greater quantities of water that could be supplied by feeder canals. Always one to combine practicality with vision, however, Chesbrough returned to his premise that a "thorough cut for a steamboat canal, to the Illinois river, which the demands of commerce are calling more and more loudly for, if ever constructed, would give . . . perfect relief to Chicago" by providing "a constant and abundant stream from the lake, flowing westwardly throughout the season of navigation, and consequently during the warm and sickly portion of the year."[5] This idea of deepening the Illinois and Michigan Canal to a depth that would carry Chicago's sewage away from the city and the lake would persist for decades to come, and eventually it would provide the template for the building of the Sanitary and Ship Canal and, in the early twentieth century, the final reversal of the Chicago River.

In the mid-nineteenth century, the cost of such an undertaking could not be fully realized, but in 1860 the Board of Sewerage Commissioners nevertheless reiterated the idea that the expense of constructing flushing canals from the lake to the river "would be rendered in large part useless, so soon as the deep cut of the Illinois and Michigan Canal be made."[6] To that end, the board on February 7, 1860, charged Chesbrough with studying two proposals, "either to cut down the summit of the Illinois and Michigan Canal, or to make a new channel through Mud Lake to the Des Plaines River, and then to deepen the channel of that stream."[7]

Chesbrough accordingly met with William Gooding, formerly construction engineer of the Illinois and Michigan Canal, who by 1860 had become secretary of the Canal Board. Gooding estimated the cost of sufficiently deepening the canal at $2 million, if the work were to be done in both summer and winter. He offered no opinion on the cost of the Des Plaines River option, but he did think that deepening the river's channel was likely to prove impossible, given the rocky nature of the riverbed.[8]

Chesbrough also reviewed the broader possibilities of creating a larger, deeper canal and of making accompanying improvements for navigation on the Illinois River, both of which would provide for steamboat traffic between Chicago and St. Louis, and he noted that "the demands of this city and State for greater facilities of transporting the grain, lumber and other products of this region, a generation hence, if the future development of our country is to be at all in proportion to the past, must eventually cause such a work to be commenced."[9] He was aware, however, that the city could not afford such an undertaking.

After reviewing the various options, Chesbrough reached the same conclusion he had put forward in his 1858 report. He believed that the condition of the river was not offensive enough to warrant the expense of building canals to flush the river's north and south branches. It was also his opinion that the river could be sufficiently cleansed during the warm months by water pumped from the south branch into the canal via the pumping station at Bridgeport. In addition, he remained convinced that the Illinois and Michigan Canal would eventually be deepened, thus obviating any flushing of the canals or any other introduction of water from the lake into the south branch.[10] Although he was correct in the latter assumption, it would take years for the deepening of the canal to become a reality.

Even as Chesbrough was laying the groundwork for addressing the city's sewage problem through the reversal of the Chicago River, a project eventually accomplished in the early twentieth century, he was called upon to ponder the second horn of Chicago's ongoing dilemma—how to provide a stable supply of pure water for the city's residents. In its January 1861 report, the Board of Water Commissioners summarized the continuing complaints about the quality of the city's water supply, pointing to "a turbid appearance to the water when the stream or lake affording the supply is agitated, an unpleasant taste to the water two or three times in the course of the year, and the periodical presence of fish in the pipes."[11] Given this state of affairs, on March 7, 1860—just a month after Chesbrough received his assignment from the Board of Sewerage Commissioners to study the canal issue—the Board of Water Commissioners passed a resolution requesting that he submit plans and an estimate for a project to extend an inlet pipe far enough out into the lake for the city's water to be free of the wash of the lakeshore and the polluting effects of the river. He was also asked to report on the feasibility of erecting any additional pumping works that might be needed to secure a supply of pure water.[12]

And so, just like that, Chesbrough became the central figure in answering Chicago's critical question of how to rid itself of its waste while also obtaining a stable supply of fresh water. The magnitude of the problem would have been enough to dissuade many another engineer, but Chesbrough brought his usual focus and careful, systematic approach to bear.

He began his analysis by asking how much water the city needed. The growth of the city had consistently outstripped estimates, and, as Chesbrough summarized, "It is very evident that what was more than enough three years ago, is totally inadequate now; and there is every reason to believe that the present supply will soon prove insufficient."[13] At the same time, he cautioned against trying to supply too great an amount of water, given the cost of constructing new waterworks. He provided the water commissioners with data showing the growth in water supply for Boston, Cincinnati, Buffalo, Cleveland, and Detroit and comparing those data to the data for Chicago. For 1858, the daily water supply for Chicago was just under 3 million gallons; for 1859, it was more than 3.8 million gallons; and for 1860, it was nearly 4.7 million gallons.[14] Although much of this growth could be attributed to increases in population, Chesbrough also noted that other causes included "the growing use of water closets and private baths, the greater watering of streets, the flushing of sewers, the increase of manufacturing establishments, and the playing of public and private fountains."[15] In view of these other variables, Chesbrough concluded that there was no consistent rule for using population figures alone to estimate future needs for water.

Another major concern in determining the amount of water needed by the city was the amount of water lost through wastage. Pointing to the experience of London, where water was supplied by private contractors who obviously had a strong pecuniary interest in limiting wastage, Chesbrough cited a report made to the Institution of Civil Engineers in March 1859 that at least two-thirds of London's water supply was being lost, much of it because of leaky mains. He then noted that while the sewerage system was being constructed in Chicago, a number of water mains had been excavated, and many leaks had been found.[16]

In 1860, the average daily water consumption for Chicago was nearly 5 million gallons, but in July of that year it had sometimes reached nearly 7 million gallons. Chesbrough proposed doubling those numbers and using 10 to 14 million gallons as the basis for estimating the daily need to be supplied by any new waterworks. With that estimate in mind, he turned to the proposal to extend a 5-foot iron pipe 1 mile into the lake. He agreed that such a pipe would be sufficient to supply the targeted amount of water, but he disagreed on the cost estimate of $65,791.20 that had been offered by Commissioner R. J. Hamilton, and he suggested instead a more realistic cost of $123,200. In addition to the high cost of such a large pipe, Chesbrough was concerned with the feasibility of placing it on the lake bed at all, and with the possibility, even if the pipe could be laid, of its becoming damaged by the dragging of anchors or by other maritime mishaps.

Given these concerns, Chesbrough turned to another option and suggested that "a *tunnel* might be made."[17] He pointed out that the onshore clay substrate likely also extended under the lake, and that if it did, substituting a tunnel for the pipe would be feasible. He suggested a plan with three shafts—one shaft at each end, and one shaft in the middle from which the work would be accomplished—and estimated that the tunnel could be completed within thirteen months after the shafts were sunk. At the eastern end, where the water inlet would be, he proposed an immovable hollow crib to protect the outer shaft. The middle shaft would be protected in the same way. As for the tunnel itself, he suggested a circular shape with a clear diameter of 6 feet, and the lining of the tunnel with brickwork 12 inches thick. An inner ring of brick would be 4 inches thick and laid in pure cement, whereas the rest of the brickwork would be laid in mortar made of equal parts cement and sand.[18]

Chesbrough estimated the cost of such a 1-mile tunnel at $126,000, but in his mind the question remained of whether either a pipe or a tunnel of such length would be sufficient to attain "the full object the Commissioners have in view."[19] He also questioned whether water obtained from 1 mile out would be appreciably better than water obtained closer to shore, especially given the expense of its acquisition. In what would prove to be a mistaken assumption, he reported that the impurities coming from the river "would be more likely to affect the water 1 mile out from the pumping station than near shore, because the prevailing winds are from the northeast, and the currents of this side of the lake strongly southward, as the character of the newly formed beach north of the pier shows."[20]

Chesbrough then went on to address three other alternatives to a pipe or a tunnel. The first was to erect new pumping works 16 miles north of the city. Although that location would still be affected by turbidity during rough weather, it would be largely removed from the influence of the impurities flowing from the river. This location, however, would require two lines of 3-foot pipes for the 16-mile run to the city, at a cost of nearly $1.7 million. He argued that two lines of pipe would be needed because the city could not risk being cut off from its water supply should a single line fail.[21] The second and third options were to construct filter beds or subsiding beds, respectively (filter beds use layers of differing materials, such as sand, shells, and gravel, to filter water; subsiding beds rely on gravity to allow sediments held in suspension to settle out over time). Although Chesbrough considered these types of beds feasible from the standpoint of cost, he thought the sheer size of filter beds would be problematic, and that difficulties would arise from their use in the harsh, freezing conditions of Chicago winters. As for subsiding beds, they would provide relatively clearer if not cleaner water but would not address the problem of the water's being affected by smoke, which gave the water "an unpleasant kreosote taste."[22]

In the final analysis, Chesbrough simply felt that he was not in a position to make a firm recommendation. He considered the pipe and tunnel options uncertain in terms of their cost, and he doubted that they would meet the board's desires in any event. As we have just seen, he believed that the filter beds would be too expensive and of doubtful efficiency in Chicago's winters, and that the subsiding beds would be too susceptible to smoke. He concluded that more time should be taken to consider the entire subject, and that more detailed sampling should be done at various distances from shore to determine the relative purity of the water.[23] And so, in early 1861, the question of how best to supply Chicago with pure water remained open, with no firm solution at hand.

On May 6, 1861, the city took its next major step regarding its water supply by combining the Board of Sewerage Commissioners with the Board of Water Commissioners to form the Board of Public Works. The new board's members were John Gindele, Frederick Letz, and Benjamin Carpenter. Chesbrough, named chief engineer, was officially put in charge of addressing the city's overlapping concerns of disposing of its waste and supplying its residents with clean water.

This consolidation of the two boards could not have come at a better time. During the winter of 1861–62, the quality of the water supply was especially bad, and the "water question" became a major topic of concern for the city. In February 1862, the *Chicago Tribune* warned that unless a supply of pure water could be found, there would be a "risk of making our town uninhabitable." The paper went on to recount the usual suspects behind the water's poor quality—runoff, and the tiny fish that tended to find their way into the water supply—but then insisted that the major cause of difficulty lay "in the floods of blood and filth and the hundreds of tons of other animal matter from the slaughter-houses, the glue factories, the packing-houses . . . the wash and slops of distilleries and other manufacturing establishments, and in the sewers which underlie the streets." These polluting sources, the paper said, fouled the river, which in turn carried the pollutants into the lake, where they mixed with the lake water to some extent before being taken up by the pumps of the waterworks and delivered "into every coffee-pot and drinking cup in the city."[24]

The Board of Public Works issued its first annual report, dated January 1, 1862, to the Common Council of the city in March 1862. In that report, the board acknowledged the growing dissatisfaction over the quality of the water but again resisted recommending any formal action. It noted that small fish often emerged from the city's faucets and hydrants but stated that the "fish nuisance" would be addressed the following spring, through additional filtration. The board then echoed the *Chicago Tribune's* charge that the larger question was one of industrial pollution, and that the primary cause of the poor water quality was "the large quantities of blood and other waste and foul substances that find their way into the river from the many slaughtering and rendering establishments, distilleries and other works of a like nature."[25]

The board concluded that if these noxious substances were stopped from being discharged into the river, the greatest source of the water's contamination would be eliminated. Barring that, however, it reiterated the five alternatives for addressing the "water question" as (1) extending the inlet pipe 1 mile into the lake (at a cost of $66,000 to $125,000), (2) constructing a brick tunnel extending 1 mile into the lake (at a cost of $126,000), (3) moving the pumping works 16 miles north of the city (at a cost of nearly $1.7

million), (4) building filtering beds (at a cost of $107,500), and (5) building a subsiding reservoir (at a cost of $107,775). The board also noted that the city still had the option of increasing the volume of water pumped from the south branch of the river into the canal at Bridgeport for more days during the year (at a cost of $150 per day) so as to temporarily improve the quality of the water and purify the river.[26] Of these five possible alternatives, the board believed that the only feasible ones were the building of filtration beds and the extension of the inlet pipe. The board made clear, however, that it saw neither of the two as desirable.

Chesbrough, in a report included in that first annual report of the Board of Public Works, reiterated and strengthened the board's view. He accepted the premise that the sewers themselves were partially contributing to the impurity of the river, and hence of the water supply, but he insisted that the broader blame lay with "the wholesale discharge of blood, and other refuse from the slaughter and packing houses, in and around the city, besides distilleries, glue factories, establishments for rendering offal, &c."[27] To prove his case, he pointed to the fact that there had been few complaints about the river during the spring, summer, and fall of 1861, even though the sewers had been flowing into it, but he noted that when the meatpacking season began, disagreeable odors had been found to affect locations on the south branch that were above the outlets of any of the sewers. Similarly, he pointed to an occasion in 1860 when the north branch had been fouled by discharges from piggeries and cow stables that were quite separate from the sewers.[28]

Chesbrough also cited the harsh winter conditions as contributing to the problem. During periods when the river was not frozen, he argued, the water was purified by its contact with the air, and this purifying effect was augmented when the water was agitated by the passage of tugs and other vessels. But during the winter, he said, ice prevented the water from coming into contact with the air, and the river then became "like a bottle with two mouths, which originally filled with comparatively pure water, but receiving only continual accessions of filth into one mouth, soon ceases to discharge anything but filth from the other."[29]

A number of water samples taken from the lake and the river had been analyzed during 1861, but Chesbrough still felt that he lacked sufficient data to recommend a permanent solution. He thought it would be premature to enact any of the five alternatives considered by the board, and so he recommended instead the temporary "expedient" of increasing the water pumped at Bridgeport to cleanse the river, as necessary.[30] There were still fundamental questions to be asked, he said, before any permanent solution could be reasonably advanced. He believed that the magnitude of the city's growth, the direction of its growth, and any possible harbor improvements needed to be considered along with the possibility of constructing a ship canal to the Illinois River. As a result, he once again demurred, choosing not to recommend a firm course and suggesting that the "light of a few years" would provide the additional data needed to properly frame future plans for addressing the city's water supply.[31]

What a difference a year would make! If in 1862, the date of its first annual report, the Board of Public Works resisted making any specific recommendations on addressing the city's need for pure water, that reluctance had disappeared by 1863, when the board issued its second annual report. In the spring of 1862, after a harsh winter left the river stagnant and full of accumulated waste, a "freshet," or quick thaw, had discharged an inordinate amount of filth from the river into the lake. With the addition of a southwesterly wind, the quality of the lake water all along the shore had become seriously compromised from at least 5 miles north to 3 miles south of the river. This event, combined with an especially bad recurrence of the "fish nuisance," was apparently the last straw for the citizenry and for the Board of Public Works.

After a quick review of the various plans that had previously been suggested and evaluated, the board finally settled on a course of action, saying that "careful investigation of the subject has satisfied us sufficiently to say, that with our present knowledge, we consider it practicable to extend a tunnel of five feet diameter the required distance under the bed of the lake, the mouth or inlet to such a conduit being the outmost shaft, protected by a pier, which will be used in the construction of the tunnel."[32] The board placed the estimated cost of the tunnel—not including a lighthouse, should one be required—at $307,552. For details of the plan, the board referred the public to the enclosed report by Ellis Chesbrough, newly appointed in 1863 to the new position of city engineer.

In his enclosed report, Chesbrough wrote that the "evil" where the water was concerned had "become such as to demand some remedy, as soon as an adequate one can be supplied."[33] After years of considering various possible remedies, he concluded that the only two plausible solutions were to take water from the lake 2 miles east of the pumping station or to take water from a point 5 miles north of the city. Either of the two, he said, would place the inlet for the water supply far enough from the river's impurities to ensure clean water for the city.

He acknowledged that his opinion in 1861—that impurities would be more likely 1 mile offshore than nearer to shore—had been incorrect, and he concluded that more recent experience showed that discharges from the river "mingle so completely with the lake as to be scarcely traceable a mile out."[34] He briefly reviewed the economic and technical difficulties of trying to install an iron pipe at such a distance, and then he turned to his preferred solution, a lake tunnel.

Chesbrough noted that borings made along the lakeshore at the Lill & Diversey Brewery showed a formation of blue clay at about 20 feet below the surface and extending 100 feet farther down before hitting bedrock. Assuming that this clay was continuous and free from beds of sand, Chesbrough stated his belief that "it would be very easy to make a tunnel large enough to supply the city for many years."[35]

He proposed that to make such a tunnel it would be necessary to sink four shafts into the lake. Each of these shafts would be protected by an octagonal crib 80 feet in diameter, with a central space 30 feet in diameter, and with approximately 25 feet of thickness left to the crib around the interior shaft. In the central space, an iron cylinder 9 feet in diameter would be sunk to the depth at which the horizontal tunnel would begin. The shaft at the eastern end of the tunnel would be equipped with a water inlet, whereas the intermediate shafts would be only for construction purposes and would be removed once the tunnel was completed.

Although in theory the tunnel would be "easy to make," Chesbrough saw fit to address two objections that he felt were likely to arise from his proposal. The first had to do with the very idea of a tunnel. Pointing to complications that had plagued the Thames tunnel in London, problems that were due to the nature of the ground, Chesbrough argued that the smaller diameter of his proposed tunnel, and the consistency of the blue clay that he expected to encounter, were enough to safeguard his tunnel from the risk of such complications. The second objection, which he took more seriously, had to do with the placement of the proposed cribs. The city had already experienced significant difficulties in its efforts to place inlet pipes for the waterworks into the lake; indeed, one of Chesbrough's primary reasons for rejecting the idea of laying a longer inlet pipe had been the potential for rough conditions on the lake. Nevertheless, he believed that properly constructed cribs could be successfully launched, towed, and sunk into place. He argued that the process would actually be easier in deeper water 2 miles from shore than it would be closer to shore, where the wave action produced more horizontal motion.[36]

Six years later, looking back at this critical time in the genesis of the plan for a lake tunnel, Chesbrough would admit to some foot-dragging by the city on the whole subject of the water supply. Despite public agitation in the press, the city had for years persisted "in the hope that much of the complaint against the water supply might be imaginary."[37] In addition, the reluctance to expend resources on the various solutions then under consideration had doubtless been coupled with the recognition that a deep cut in the Illinois and Michigan Canal would have an immediate cleansing effect on the Chicago River and would thus eliminate much of the need for an offshore inlet for the city's water supply. The city's reluctance had finally given way, however, when "in the course of two or three years more, the water supply became occasionally very offensive, both to the taste and the smell." At that point, Chesbrough recalled, the city had finally concluded that a "remedy for this state of things could no longer be neglected."[38]

After "much doubt and careful examination of the whole subject," Chesbrough went on, he had become convinced that a tunnel was the best solution because it combined "greater directness to the nearest inexhaustible supply of pure water, with permanency of structure and ease of maintenance."[39] Others had not shared his confidence, however, and he had duly acknowledged "the possibility, and in the estimation of many, the probability, of meeting insuperable difficulties in the nature of the soil, or storms, or ice on the lake."[40] To counteract these doubts, he had addressed them "one by one" and found ways to overcome them, "either by providing against them, or discovering that they had no real foundation."[41]

In making his case for the lake tunnel, Chesbrough had turned to colleagues of great reputation and skill. He submitted the early plan to a number of engineers, including Colonel James D. Graham, superintending engineer of the Chicago lighthouse; John B. Jervis, his friend and colleague from his days in Boston; Captain William H. Swift, one of the first and most respected engineers to have emerged from the US Military Academy, and a longtime colleague of Colonel Stephen H. Long; and Leverett. H. Clark, chief engineer of the Illinois and Central Railroad and later chief engineer for the City of Chicago. These men, imbued with the expertise and traditions of both the US Military Academy and so-called Erie School of Engineering, represented a cross-section of the highest level of American engineering. Chesbrough was understandably pleased that all of them had "expressed their belief in the practicability of the scheme."[42]

As is true of all public works projects, especially those of novel design that challenge conventional wisdom, the engineering aspects of Chesbrough's plan constituted only one aspect of the problems to be overcome. Politicians were to make the final decision about whether to proceed, and they had to be convinced both of the plan's probable success and of its cost efficiencies. To address the latter issue, Chesbrough, in his contribution to the 1863 report of the Board of Public Works, reviewed the most applicable costs available. Once again, the Thames tunnel proved a good starting point, and Chesbrough made good use of the fact that he expected the lake tunnel to be far more cost-effective, given the anticipated soil conditions. He made a similar argument for the excavation costs, modeling them on the sewers of London while again discounting the cost because of the favorable soil conditions in Chicago. For the sinking of the shafts, he cited the Theiss Bridge in Hungary, the railroad bridge across the Savannah River, and the Harlem Bridge in the United States. For masonry, he turned to the Boston Water Works, and for the crib costs he relied on reports from Colonel Graham for prior work done along the Chicago lakefront and in Michigan City. Finally, he compared his cost estimates to other tunnel projects, relying on reports from the commissioners of the Troy and Greenfield Railroad and the Hoosac Tunnel in Massachusetts, and he included an analysis of twenty tunnels that had been built in England and France. Having presented this broad array of data, he stated his hope that the breadth and depth of his cost analysis might "afford some assurance" for the commissioners of his own Board of Public Works.[43] He then summarized his cost estimate for the lake tunnel, leaving out any lighthouses that might be erected on the proposed cribs, as follows:

16,000 cubic yards excavation at $3 per cubic yard = $48,000
4,750 cubic yards masonry (including land shaft) at $20 per cubic yard = $95,000
561,600 cubic feet of cribs at 22 cents per cubic foot = $123,000
Cylinders 210 lineal feet, and apparatus for and sinking of shafts = $41,000
TOTAL = $307,552

The relatively low cost that Chesbrough estimated for the project was one that occasioned mixed views. On the one hand, the projected cost compared very favorably with the likely costs of the other alternatives that the commissioners had considered. On the other hand, as Chesbrough would recall in 1868, it was an estimate "in which not only the public generally, but the [members of the] Board of Public Works themselves, had no great confidence."[44]

Chesbrough was supported early on by Benjamin Carpenter, the first president of the Board of Public Works. William B. Ogden, Chicago's first mayor and a man of considerable influence, also offered to support the project publicly, should it become necessary. Chesbrough initially received a less favorable hearing from John Gindele and Frederick Letz, the board's other members, although they finally came to offer positive if cautious support. A new board was chosen in 1863, replacing Carpenter with Orrin J. Rose, and, what was more significant for the tunnel plan, adding Mayor Francis C. Sherman as an ex officio member. Sherman, whose private experience was in brickmaking and construction, would prove to be a great ally to the plan. At his suggestion, further borings were made at 500-foot intervals to confirm Chesbrough's belief that a sufficiently thick layer of blue clay extended from the northeastern portion of the city out into the lake and to the proposed location of the crib at the end of the tunnel.

It was already known, from borings previously taken at the Lill & Diversey Brewery, that a blue clay formation began at about 20 feet below the ground's surface and extended to a depth of 100 feet or more. On June 11, 1863, Chesbrough began taking borings from the lakebed itself. He employed Captain A. S. Bergh as foreman, and Berg was assisted by C. W. Fuller, Charles Simmons, P. H. Crosby, George Hall, and John Igo. To take the borings, two scows were towed into location and securely anchored. Between the two boats, a 2-inch gas pipe was lowered, the top of the pipe being 2 or 3 feet above the surface. An auger was then passed down the pipe, with sections of pipe and auger being added as necessary to reach the desired depth. Both the pipe and the auger were raised and lowered with a 25-foot-high derrick, by means of block and tackle. The first boring was taken 0.75 mile from shore, where the water was 28 feet deep. Beneath 4 inches of sand, the crew penetrated 30 feet and found nothing but blue clay. The second boring was 1.75 mile out, in 30 feet of water, and it yielded the same result. The last boring was 2.25 miles out, in 36 feet of water. Here the lakebed was found to be covered by 1 foot of mixed sand and soft clay, beneath which the clay became increasingly thick. After penetrating 6 to 8 feet, the crew found clay "of a bluish slate color, of very fine grain, with little or no grit [that] would probable [*sic*] make excellent brick." Overall, the first three borings suggested clay of similar character at 30 feet below the lakebed over the proposed 2-mile track of the tunnel.[45] This result was borne out by the next three borings.

On June 16, Chesbrough escorted more than twenty citizens, along with Board of Public Works members Gindele, Letz, and Rose and J. K. Thompson, Chicago's superintendent of streets and bridges, aboard the tugboat *G. W. Wood* to witness the seventh boring experiment. The boring, made 2 miles out, produced the same result as the prior six. The board members were satisfied enough by these preliminary findings to authorize the preparation of plans and specifications for potential bids. It was also determined, in response to Mayor Sherman's wishes, that additional borings would continue to be made so that the character of the earth would be known every 500 feet along the track of the proposed tunnel. For the additional borings, however, a different method would be used, one that employed a valve at the bottom of the drill pipe so that when the pipe was drawn out it would contain a core sample of the boring. The board members were understandably encouraged by the findings of the seventh boring, perhaps in part because on the day they accompanied Chesbrough the water 2 miles out was 36 feet deep and "so clear that an object the size of a man's hand, suspended by a cord, was seen at a depth of eighteen feet." At three o'clock on that afternoon, the temperature of the surface water was 60 degrees Fahrenheit, whereas at the bottom of the lake the temperature was 51.5 degrees.[46]

Some weeks later, the *Chicago Tribune* reported that the plans and specifications for the tunnel project were nearly completed, and the paper anticipated that contracts for various aspects of the plan "would be let at an early day." In describing the plan, the paper termed it "a gigantic undertaking, the first of the kind ever attempted on this continent."[47] The following week, the paper published a detailed explanation, including drawings, of the proposed plan, and on the same day the paper also published a letter by a "Tax Payer," who lamented, "I have little doubt but that the bill [for the tunnel] will be millions where these inquiries have given us hundreds of thousands."[48] The anonymous writer went on to declare the project an obstruction to navigation and to challenge its very feasibility. This letter was but the first intimation of a growing wave of resistance to the project.

By mid-August of 1863, the project was ready to be bid. In the last two weeks of that month, the following notice appeared in newspapers in Chicago, New York, and Boston:

> Proposals for Constructing a Brick Tunnel under Lake Michigan two miles long, for the City of Chicago.
>
> Office of the Board of Public Works, Chicago, August 13th, 1863
>
> Sealed Proposals will be received at this office until WEDNESDAY September 9th, at 11 o'clock A.M at which time the Board will open the same, for doing all the work and furnishing all the materials required for the construction of a Brick Tunnel, five feet clear diameter, and extending from the present pumping works of the city, two miles out, under the bed of Lake Michigan, and through what numerous borings invariably show to be a stiff blue clay soil, together with land and lake shafts, and protecting cribs, in accordance with the plans and specifications for the doing of said work, to be found on file in the office of this Board on and after the 19th inst.
>
> The Board would prefer to let the entire work to one contractor or company, but will receive separate proposals for the tunnel proper and land shafts, for the protecting cribs, and for the cast iron cylinders for lake shafts.
>
> The Board reserves the right to reject any or all bids, and to accept any one bid complying with the conditions of this advertisement.
>
> The bids must be sealed, and must be accompanied with a bond of the penal amount of $200, blanks for which will accompany the specifications, guaranteeing the execution of a contract in accordance with the bid, in the case the bid is accepted.
>
> Proposals must be directed to the Board of Public Works, and indorsed "Proposals for Lake Tunnel."
>
> Satisfactory security for the faithful performance of the work will be required of the parties whose bids may be accepted.[49]

The proposal was signed by Board of Public Works members Gindele and Letz and by Mayor Sherman.

"The opening of the proposals was looked to with great interest," Chesbrough would later recall, "as it was feared that no responsible parties would offer to take the work for less than millions instead of only about three hundred thousand dollars."[50] Despite this trepidation, when the bids were opened, on September 9, 1863, they were both "satisfying and gratifying."[51] The bids ranged from $239,548 to just over $1 million. The two lowest bids were conditional and were therefore rejected as uncertain. That left the unconditional bid of $315,139 of James J. Dull and James Gowen (Dull & Gowen) of Harrisburg, Pennsylvania, as the lowest and best bid. The board accepted this bid the next day, encouraged by work that Dull & Gowen had done on the Croton Water Works and in New York's Central Park.[52] The board recommended that the Common Council accept the contract and requested authorization to proceed with the requisite bond issuance that would fund the project. The *Chicago Tribune* reported favorably on the board's acceptance

Bids Received for the Lake Tunnel, September 9, 1863.

NAMES OF BIDDERS.	Lake Tunnel complete	Lake Tunnel and land shaft alone in case there should be two (2) lake shafts.	Lake Tunnel and land shaft alone in case there should be three (3) lake shafts.	Lake Tunnel and land shaft alone in case there should be four (4) lake shafts.	Outer Crib.	One Intermediate Crib.	2d Intermediate Crib.	3d Intermediate Crib.	Each additional foot in height.	Outer Cast Iron Cylinder.	First Intermediate Iron Cylinder.	Second Intermediate Iron Cylinder.	Third Intermediate Iron Cylinder.	Each lb. of increased length or size.
James Andrews, ... (Pittsburgh, Pa.)	$239,548	$151,987	$148,000	$144,000		$26,021	$23,419	$20,000	$800.00	$14,320*	$11,112	$11,112	$11,112	4 1-5 cts.
James J. Dull & James Gowen, (Harrisburg, Pa.)	$315,139													
Stephen C. Walker, Asa D. Wood, & F W. Robinson, ... (New York.)	$315,000†								$200.00					10 cts.
Thos. Williams, John McBean, A. S. Brown, & George Neilson, (Chicago.)	$490,000‡								$1,250.00	16c ℔ lb.	16c ℔ lb.	16c ℔ lb.	16c ℔ lb.	16c ℔ lb.
Hervey Nash, ... (Chicago.)	$40 per lineal ft.													
D. L. DeGolyer, ... (Chicago.)	$620,000§	$410,000	$400,000	$380,000 "360,000?"										
William Baldwin, ... (New York.)	$100 per ‖ lineal ft. $1,056,000													

* Included in first general proposal, and would not be contracted for separate. Would deduct from my first general proposition, in case the Board furnish and sink this cylinder, $14,320.

† Meaning to include but one intermediate lake shaft. This proposal is based upon the supposition that the material to be excavated is firm, and such that the tunnel can be made and the masonry built without the use of permanent bracing. But if in the prosecution of the work, to secure its safety, permanent bracing should be required, the attendant increased expenses shall be paid by the City. The Chief Engineer to decide the amount.

‡ For all the work except the iron cylinders for shafts. In case sand or gravel veins occur, to be paid for extra. This sum includes four cribs.

§ I run the risk of all loose earth, sand, or gravel.

‖ $100 per lineal foot, if material is "stiff blue clay soil." If otherwise extra pay and extra time will be required.

Figure 7. Bids on the lake tunnel. Source: City of Chicago, *Eighth Annual Report of the Board of Public Works to the Common Council of the City of Chicago for the Municipal Fiscal Year Ending March 31st, 1869* (Chicago: Jameson and Morse, 1869).

of Dull & Gowen's bid and went on to "congratulate the Board and the public, at the present fair prospect of an early completion of the Tunnel, and of securing to the city thereafter an abundant supply of pure water."[53]

On September 11, Chesbrough sent a note to the Board of Public Works suggesting that an additional $60,000 be added to the cost of the project to cover "additions to the cribs, lighthouses; connections with the well of the pumping works; engineering, superintendence, and contingencies, making the whole cost $375,139."[54] That evening, the Common Council held a special meeting to discuss the board's report and its recommendation that the council accept the Dull & Gowen bid. Present at the meeting were Mayor Sherman and Aldermen Hahn, Barrett, Kahn, Sheridan, Walsh, McDonald, Abbott, Comiskey, Brown, Himrod, Bond, Gastfield, Aleckner, Hottinger, Sullivan, Shufeldt, and Woodman. After the mayor read the board's communication, the Common Council had a vigorous debate. Some aldermen wanted a special committee appointed to review the board's recommendation and the contract, whereas others felt that the matter had been sufficiently vetted by the board. Alderman Comiskey was forthright in saying, "This contract is a big thing, and I am afraid to undertake it." After much discussion, a special committee of five, comprising the mayor and the comptroller as well as Aldermen Hahn, Brown, and Woodman, was appointed and asked to report back to the Common Council at a special meeting to be held four days hence.[55] As the current Common Council meeting unfolded, it must have appeared to Chesbrough that there was more quicksand to be feared than that found underground.

The special committee, some of whose members "were known to have strong doubts with regard to the propriety of undertaking such a work,"[56] met over the following four days and ended up giving the project, and the selection of Dull & Gowen, its strong blessing. At the special meeting held on September 15, 1863, the committee stated, "After careful examination and inquiry of scientific and practical men, we are satisfied that the project is judicious and sound, and will prove a perfect success. That it is novel and extraordinary, and will be regarded as one of the wonders of the age, only entitles its originators to greater credit." The committee went on to affirm the choice of Dull & Gowen as the lowest and best bidder for the contract.[57]

At the same meeting, however, another obstacle arose. James Andrews, the nominal low bidder for the project, had sent the Board of Public Works a long communication objecting to the board's selection of Dull & Gowen and asking to have his bid reconsidered by the Common Council. Andrews claimed that his own bid was the lowest and best, and he threatened to pursue legal action if it were not accepted. A motion was made to refer this communication to the special committee, and after a "spirited debate" the reference was made on a vote of eleven ayes to six nays.[58] The special committee subsequently reviewed and rejected the petition by Andrews, concluding that the Board of Public Works had been correct in its selection of Dull & Gowen.

At the next meeting of the Common Council, on September 21, the tunnel nevertheless remained an unsettled issue. First a self-styled civil engineer named James Hamilton urged the council not to ratify a contract for the tunnel and instead to accept his idea of filtering water through sand along the lakeshore. Alderman Shufeldt found the idea appealing, stating that it had taken thirty years to complete the Thames tunnel and that it was only 1,000 yards from bank to bank (in fact, however, the length of the Thames tunnel is 1,300 feet). The council voted to refer Hamilton's proposal to the Board of Public Works. Also at that meeting, a communication from Thomas Williams, John McBean, George Neilson, and A. S. Brown was read regarding their rejected bid for the tunnel. They argued for a tunnel 6.5 feet in diameter and said they could construct it for about the same cost as the 5-foot-diameter tunnel. Their communication was also referred to the Board of Public Works. Even more troubling for the tunnel project, a motion to approve the board's acceptance of the tunnel contract failed by a narrow vote. The thirteen ayes were Aldermen Hahn, Schall, Kahn, Sheridan, Brown, Himrod, Holden, Von Hollen, Bond, Gastfield, Ruh, Hottinger, and Woodman; the fourteen nays were Aldermen Shimp, Titsworth, Barrett, Gallup, Walsh, McDonald, Comiskey, Ullbrich, Casselman, Armstrong, Aleckner, Conlan, Sullivan, and Shufeldt.[59] Obviously, there were more hurdles in store for the Chesbrough Plan before spade could be put to soil.

Things came to a head at the Common Council meeting of October 5, 1863. A communication was presented by L. P. Hilliard and others, asserting that the proposed tunnel would be "a failure . . . merely designed to put money in the pockets of its projectors" and asking that the project be stopped. The council referred this petition to the Committee on Harbors and Bridges. Next, Chesbrough reported on his review of the plan previously proposed on September 21 by James Hamilton. Chesbrough estimated that Hamilton's plan would cost $636,620, that it would take at least two years to build, that it would be "utterly inadequate" to meet the city's present and future needs, and that, in any event, it would not satisfactorily purify the water, contrary to what Hamilton had asserted. Chesbrough also replied to the communication of Messrs. Williams, McBean, Neilson, and Brown, noting that the larger tunnel they suggested had already been allowed for in the proposal by Dull & Gowen, and at a lower cost.[60]

Alderman Barrett then moved to reconsider the council's vote of September 21, which had rejected the special committee's recommendation to approve the Board of Public Works's action on the tunnel and accept the contract with Dull & Gowen. Alderman Conley attempted to table Barrett's motion, but Conley's motion failed by a vote of eighteen to twelve. The motion to reconsider then passed by a vote of eighteen to twelve, placing before the council the question of adopting the board's action and passing the requisite ordinance. Alderman Comiskey argued that the bid by James Andrews should be accepted because it was the lowest one. Debate ensued concerning the petitioner's qualifications, his level of responsibility, and his character, with Alderman Comiskey claiming to have been informed that there were sufficient testimonials on these issues. The mayor countered that when the special committee endorsed the action of the Board of Public Works, the only testimonial available for James Andrews had been one from Andrews himself. Alderman Walsh took a different tack, arguing that the tunnel project was unnecessary because flushing the river with water from the Illinois and Michigan Canal would resolve the issues with water quality. He also asserted that the contract as written allowed for the possibility of additional charges and work. Finally, he argued, the crib would not last—sand would wash out from beneath it, and it would topple over.

Alderman Brown countered the points made by Comiskey and Walsh. He rejected any suggestion of corruption in the selection of Dull & Gowen, saying that he was not anxious for them to receive the contract but was anxious for pure water. If Walsh was satisfied with the quality of the water, said Brown, then Walsh had not so good a faculty of taste as did Brown. As for the possibility of extra work under the contract, Brown pointed out that Walsh had glossed over the fact that such work would be allowed only if it had been omitted from the specifications.

Walsh made one more attempt to delay the vote on adopting the recommendation of the Board of Public Works to accept Dull & Gowen's bid and propose the requisite ordinance, but the mayor said Walsh's motion was too late, and he called the question. The motion to adopt the board's recommendation and accept the contract carried by a vote of eighteen to twelve. The ayes were Aldermen Hahn, Schall, Titsworth, Barrett, Gallup, Edwards, Kahn, Sheridan, Abbott, Brown, Himrod, Holden, Von Hollen, Bond, Gastfield, Ruh, Hollinger, and Woodman. The nays were Aldermen Walsh, McDonald, Comiskey, Ullbrich, Clark, Talcott, Armstrong, Aleckner, Conlan, Casselman, Sullivan, and Schufeldt. By the same vote, the Common Council passed the following ordinance:

> Section 1. Be it ordained by the Common Council of the City of Chicago, That the board of public works are hereby authorized to extend the inlet pipe of the waterworks, two miles out into lake Michigan, by the construction of a brick tunnel under the bed of the lake, according to the specifications prepared by the board of public works for the same, entitled "specifications for lake tunnel," and appended herewith, and to contract for the doing of the work.
>
> Section 2. The board of public works are hereby authorized to issue one hundred and fifty (150) water loan bonds, of the denomination of one thousand dollars each, to be dated July 1st, 1863, and payable in New York July 1st, 1888, with coupons attached for interest, at the rate of seven per cent. per annum, payable semi-annually in New York, the said bonds to be issued in conformity with the provisions of an act of the general assembly of the State of Illinois, entitled "An Act in amendment of, and supplemental to, an act, entitled 'An Act to incorporate the Chicago City Hydraulic Company,' " approved February 16th, 1857; and also of an act of the general assembly of the State of Illinois, entitled "An Act to reduce the charter of the city of Chicago, and the several acts amendatory thereof, into one act, and to revise the same," approved February 13th, 1863.
>
> Section 3. The board of public works are hereby authorized to issue, in addition to the bonds above named, two hundred and twenty-five (225) water loan bonds, of the denomination of one thousand dollars each, to be dated July 1st, 1864, and payable in New York twenty-five years after the date thereof, with coupons for interest, at the rate of seven per cent. per annum, payable semi-annually in New York, the said bonds to be issued in conformity with the provisions of an act of the general assembly of the State of Illinois, entitled "An Act to reduce the charter of the city of Chicago, and the several acts amendatory thereof, into one act, and to revise the same," approved February 13th, 1863.[61]

With this authorization in hand, the Board of Public Works moved forward to finalize its agreement with Dull & Gowen. Further investigations were made into the ability and qualifications of Dull & Gowen to perform the work, and the board sent the city's legal advisor, Benjamin F. Ayer, to Harrisburg to confirm the sureties offered by contractor. Upon his positive report, the board finalized the contract for presentation to the Common Council.

The contract was supposed to be signed on October 20, 1863, but on that day another salvo was fired by the opposition. Alderman Walsh, renewing a now familiar theme of the opposition, submitted a communication to the Common Council arguing that the building of the proposed crib would be a permanent obstacle to navigation and was therefore illegal.[62] The council referred the matter to the Committee on Harbors and Bridges, and on October 24 the Board of Public Works again called on Ayer and asked him to research the question. On November 2, the Committee on Harbors and Bridges recommended that the council reject Walsh's argument, and on November 5, after searching "several cart-loads of law reports," Ayer presented his report, concluding that Walsh was mistaken and that the city had full authority to authorize construction of the tunnel and cribs.[63] From an abundance of caution, however, the council decided to ask Congress to specifically approve the proposed construction.[64]

Also on the day when the contract with Dull & Gowen was to have been signed, the *Chicago Tribune* published a further blast from the would-be civil engineer James Hamilton, wherein he claimed that the tunnel, "if ever attempted to be built, would be impracticable and a failure." He took issue with all aspects of the plan, claiming that the proposed crib would never be able to stay in place on the lakebed, that it would take ten shafts to supply enough air for the men digging the tunnel, that the tunnel would not be able to handle the water flow, that multitudes of fish would enter the system, and that the project would cost upwards of $2 million, or double his earlier estimate of $1 million. He added that if his plan were to be approved, he would be glad to build it and have it up and running by May 1864. He offered no evidence for his criticisms or his cost analysis other than to say that his "experience in tunneling [had] been considerable."[65]

Two days later, Hamilton was back with another letter to the *Chicago Tribune*, this time to renew his charges that the Chesbrough Plan would not work and that only his own plan was advisable. In this letter, Hamilton also made various claims about Lill & Diversey Brewery's digging of a well on its property.[66] In the next day's issue, Lill & Diversey conclusively refuted Hamilton's claims.[67] Five days after that, Michael McDermott, a civil engineer and surveyor, drove the final nail into Hamilton's proposed plan by showing that Hamilton's calculations were scientifically and mathematically impossible, and he concluded, "Mr. Hamilton's knowledge of his subject is erroneous, and . . . we may state without fear of contradiction, that all his other reasonings on the same subject [are] also incorrect."[68] No further communications were made by Hamilton after these responses.

Chesbrough would later comment, "About this time the public press, which had previously almost unanimously spoken favorably of the project, turned almost as unanimously against it. This opposition gradually diminished but did not cease entirely till the work was nearly completed."[69] The pages of the *Chicago Tribune* were representative of this turn of opinion. The day after Michael McDermott demolished James Hamilton's claims, the paper stated, "It would be presumptuous of us to say that we do not believe in the plan of obtaining fresh water for the city's use, which has been adopted by the Board of Public Works, and approved by the City Council" but then completely reversed its prior support of the plan and asserted, "We have the authority of the best engineering talent in the West, when we say the success of that plan is more than problematical." The *Chicago Tribune* never identified this "best engineering talent" but went on to argue that improvements to the Chicago harbor should be combined with the city's quest for pure water. It then pressed its own plan for building a pier that would protect the harbor from sediment while also protecting an extended water-intake inlet pipe, which, the paper claimed, would "answer all purposes for a quarter, perhaps half a century."[70]

The *Chicago Tribune* upped the ante the next day by denouncing both the Chesbrough Plan and the officials, elected and otherwise, associated with it. Citing an unnamed "informant"—perhaps the unnamed "best engineering talent" from the previous day's opinion—the paper urged the Common Council to suspend work on the tunnel so that a way of combining harbor improvements and water supply could be considered.[71] The paper also worried about the continuity of the clay substrate through which the tunneling was to take place, fretted about what it saw as the uncertainty of the project's cost, and once again raised the specter of the Thames tunnel, concluding, "This is no time . . . for experiments."[72]

The *Chicago Tribune*'s criticism of the plan, and of the men behind it, continued in an even more vitriolic and personal fashion when the paper called Mayor Sherman "brainless" and "incompetent" and, referring to the Common Council and Board of Public Works, asked if "anything could be more pitiable than the condition of Chicago today." The paper lamented the condition of the harbor and called the river a "seething, stinking pool." As for the plan to supply the city with fresh water, the paper wrote, "We have a project for a tunnel, difficult and uncertain of execution, costly and after all of doubtful utility—a tunnel two miles in length, compared with which the Thames tunnel is a baby."[73] Ten days later, one John Gage joined the chorus of opposition, writing that there was "a novelty and boldness to the projected tunnel" that he "admire[d]," but that, "as a taxpayer," he found that the project "reache[d] too deep down in my pockets for an uncertainty."[74]

Although, again, the contract was supposed to have been signed on October 20, it was not until November 19 that the Board of Public Works finally did sign it and agree with Dull & Gowen that the company would "construct, complete, and put in readiness for use, and keep in perfect repair for one year after the same shall have been faithfully completed to the satisfaction of the

Board of Public Works of said city, a tunnel from the lot now occupied by the pumping works of the Chicago Waterworks, and extending two miles out under Lake Michigan, with the appurtenances to the same, and to take the risk of all disturbing causes and obstacles which may occur in the construction of said tunnel and appurtenances."[75] The Chesbrough Plan, at least contractually, was at last a reality.

At the November 23 meeting of the Common Council, the Board of Public Works formally recommended that the council reject Alderman Walsh's attempt to suspend work on the tunnel until Congress authorized the erection of the cribs. The board reported that it had already executed the contract, and that the contractors had begun work. The board went on to argue that any delay would "cause great injury to the city, and postpone the supplying of the city with pure water, while no great good would be [e]ffected thereby." It noted that the city had already taken measures to obtain the sanction of Congress for the proposed work, and that the structures, if declared unlawful, could simply be removed to a depth that would prevent their hindering navigation.[76] The Common Council concurred in this recommendation, and, with this last political hurdle cleared, the Chesbrough Plan was ready to be implemented.

On November 30, water was drawn off the Illinois and Michigan Canal and into the south branch of the river, an operation that caused a discharge of filth into the lake. Shortly thereafter, southerly winds pushed the pollution northward, with resulting poor quality of the city's water supply. On December 3, perhaps because of repeated charges by the tunnel's opponents that the quality of the water would not be good even if the water were obtained from a spot 2 miles out from the lakeshore, representatives from the Board of Public Works conducted an investigation. They sampled water taken from an offshore site adjacent to the waterworks and found it to be "nearly as filthy as in the river," but water samples taken from 2 miles out, in the vicinity of the proposed crib, were found to be "perfectly pure." They then took a sample along the north shore, near what is today the intersection of Lake Shore Drive and Sheridan Road, and found the water there to be basically as foul as the first sample.[77]

But a final voice of opposition was still to be featured in the pages of the *Chicago Tribune*. In a letter to the paper, "A Tax-Payer" argued that the tunnel could "never be made at any expense," and that even if it were completed, it would not provide pure water. Repeating the now familiar arguments of those in opposition to the plan, the writer argued that if sand or gravel were to be encountered at any point in the tunneling, water would be "forced through like a stream from a fire engine, to wear a hole large enough to fill the tunnel in ten minutes, and consequently drown all the workmen." The writer concluded that if such an event occurred, the city would never be able to find workers willing to finish digging the tunnel, even if such an obstacle could be overcome.[78]

At last, however, on December 16, 1863, after a year of political and public wrangling during which there must have been times when Chesbrough wondered if his plan would ever become a reality, Dull & Gowen began physically preparing the site for the land shaft. Their first job was to erect a house over the site of the land shaft and build the requisite offices for the project. With the approval of the Common Council, and with a signed contract in hand that specified the work to be done, Dull & Gowen were missing only one final piece before they could begin in earnest. To satisfy the concerns of those who had raised the issue of the project's legality amid worries about the cribs' potential to impede navigation, the city had requested an act of Congress authorizing the erection of the cribs in the navigable waters of Lake Michigan. The final missing piece was provided early in 1864:

AN ACT DECLARING THE ASSENT OF CONGRESS TO AN ACT OF THE LEGISLATURE OF THE STATE OF ILLINOIS, THEREIN NAMED.

> *Be it enacted by the Senate and House of Representatives of the United States of America, in Congress assembled*, That the consent of the Congress is hereby given to the operation of the eleventh section, chapter fifteen, of the act of the General Assembly of the State of Illinois, approved February thirteenth, eighteen hundred and sixty-three, entitled "An act to reduce the charter of the City of Chicago, and the several acts amendatory thereof, into one act, and to revise the same," which section is as follows:
>
> "Said city shall have the power to extend aqueducts or inlet pipes into Lake Michigan, so far as may be deemed necessary to insure a supply of pure water, and to erect a pier or piers in the navigable waters of said lake, for the making, preserving and working of said pipes or aqueducts: *Provided*, that such piers shall be furnished with a beacon light, which shall be lighted at all such seasons and hours as the light on the pier at the entrance of Chicago river." APPROVED, January 16, 1864.[79]

With that, the legal and political obstacles to the building of the Chicago lake tunnel were overcome. The *i*'s were dotted and the *t*'s crossed. Although naysayers would continue to doubt the efficacy and ultimate success of the plan, the way was cleared for work to begin. It was at last time for the Chesbrough Plan to become a reality. It was time to build the great Chicago lake tunnel.

Chapter 5

✦

Building the Great Chicago Lake Tunnel

On January 13, 1864, the first of the three iron cylinders for the land shaft, all of which were manufactured by James Marshall & Company, were loaded onto railcars in Pittsburgh. Because of intervening issues with the railroad, however, the first of the cylinders did not arrive in Chicago until February 6. When it arrived, the first cylinder had to be transferred to the Galena & Chicago Union Railroad because the switch at the city could not handle the cylinder's massive size. Each section weighed 29,000 pounds and was 9 feet in internal diameter, 10 feet long, and 1.5 inches thick. Each was unloaded at the Rush Street Bridge and then rolled to the work site. The cylinders for the land shaft were so large and heavy that there was only one railroad car in the country capable of transporting them, and so it took until March 10 for all three of the cylinders to be delivered.[1] It was a considerable undertaking to transfer all three to the worksite, where the requisite office buildings and engines had been erected. But the last of the cylinders was rolled into place on March 16, the day before the groundbreaking ceremony was scheduled to take place.

The following morning, March 17, about one hundred people gathered at the worksite, at the corner of Pine Street (now Michigan Avenue) and Chicago Avenue, for the groundbreaking. Mayor Sherman was there to preside, joined by Messrs. Letz and Rose from the Board of Public Works; S. S. Hayes, the city comptroller; U. P Harris, chief of the fire department; and Aldermen Brown, Holden, Schall, Titsworth, Von Hollen, Woodman, Ruh, Kahn, Himrod, Sheridan, Sullivan, Shimp, and Gastfield. Also there to witness the official beginning of his great plan's execution was the city engineer, Ellis Chesbrough.

A little before eleven o'clock, Mayor Sherman called the ceremony to order and gave a brief speech summarizing the long history of the water question and the various remedies that had been tried and proposed. He reviewed the long, two-year process by which the Board of Public Works had examined the various plans before settling on the tunnel as the best one, and he accentuated the support the plan had received from "scientific gentlemen." The mayor also emphasized that Dull & Gowen, through their unqualified

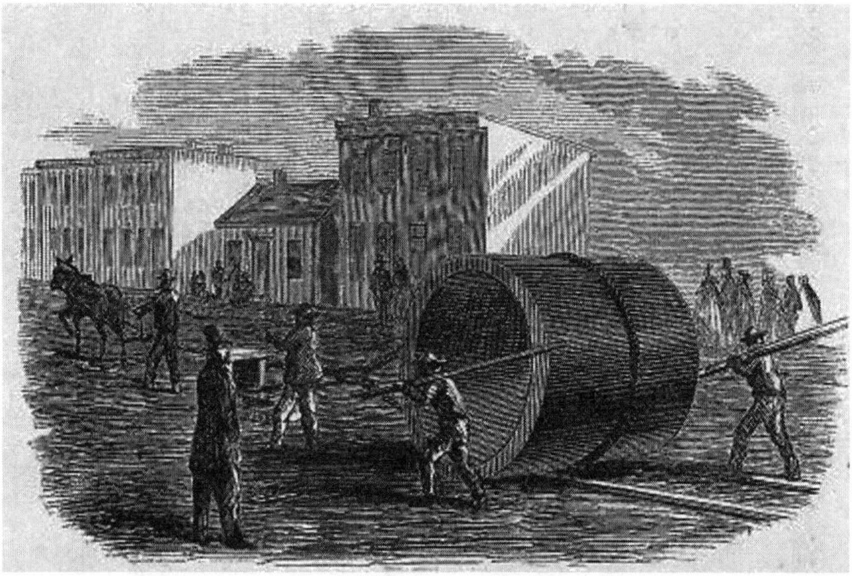

Figure 8. Delivering one of the cylinder sections for the land shaft. Source: *Harper's Weekly*, May 7, 1864.

Figure 9. Groundbreaking ceremony for the great Chicago lake tunnel, March 17, 1864. Source: *Harper's Weekly*, May 7, 1864.

Figure 10. Worksite for land shaft (note the breakwater in the lake, protecting the existing water intake). Source: Chicago History Museum, ICHi-85697, Charles R. Clark Collection. Photographer unknown.

bid, had assumed all financial responsibility for the project.[2] With that, the mayor took up a pick and broke ground amid cheers from the crowd. Each of the other assembled dignitaries then took a shovelful of dirt and placed it into a wheelbarrow, which was dutifully hauled away by James Gowen, and the ceremony was over. As a newspaper account of the event noted, "No expensive demonstration was made—not even a bottle of wine was spared to break over the ground; but all that was necessary was done, the lake tunnel was commenced."[3]

The original specifications had called for the entire land shaft to be built from brick, but, given the sand and gravel expected to be encountered for the first 20 feet or so, a decision was made to use iron cylinders for the first section before switching over to brick. The bottom of the lowest cylinder had a cutting edge, and the other sections had internal flanges where they were joined together by bolts and rust joints (a wet mixture of iron filings, sulfur, and sal ammoniac that, once oxidized, formed an impervious seal). The uppermost cylinder was designed to be fitted with an air lock so that the pneumatic method of sinking the cylinders could be used if necessary.[4] Given the relatively soft ground encountered as the shaft progressed, however, it was found that the pneumatic method was unnecessary.

Figure 11. Hoisting apparatus for the land shaft. Source: J. M. Wing, *The Tunnels and Water System of Chicago: Under the Lake and Under the River* (Chicago: J. M. Wing & Company, 1874).

The worksite itself was quite unassuming in the early days of excavation. A visitor would have seen only the office of the contractors, apart from a mound of clay and a small shed containing a steam engine. It would have been difficult for the casual observer to have guessed at the great work taking place below.

The land shaft itself was surrounded by a small enclosure, and an elevator was constructed to carry the miners up and down the shaft to their work, the elevator being propelled by the steam engine. Later, after tunneling had begun, the same elevator was used to transport brick, cement, and other materials down to the workers, and to bring the excavated earth back up.

The early days of sinking the shaft were complicated by heavy rains as well as by problems with the pumps that were used to clear water from the excavation. The pumps became clogged with sand and had to be cleaned repeatedly. By April 7, the rains had finally let up, and a steam-driven pump had been rigged to remove the surface water. After that, the miners were able to resume excavation, and by April 9 they had succeeded in digging through the stratum of sand and had hit blue clay at about 18 feet. At noon on April 11, the last of the cylinders was bolted into place, and the miners were about 3 feet into solid clay.[5]

Once the miners were into the blue clay, things began to progress apace. By April 20, they were 49 feet below ground, and the quality of the clay was thought to be so good that the contractors and the city determined that it could be used for making the bricks for the tunnel. It was thus decided to erect an on-site brickmaking facility, which was expected to be of great help in the work because of brick shortages that had been encountered early on. Unfortunately, however, this plan proved unsuccessful because the quality of the clay proved not to be as good as originally thought, and so the contractors had to procure brick from outside vendors. By April 23, the miners were at 62 feet, and by April 27 they had reached 75 feet, the full depth of the shaft needed to connect to the tunnel, including a sump 6 feet below the floor of the tunnel to collect water. The next step would be to brick up the shaft in preparation for the tunneling operation. It was anticipated that the push eastward could begin in the next day or so.[6]

Then came a near disaster. At two o'clock on April 28, a jet of poisonous flammable gas erupted in the shaft about 15 feet from the bottom, at a junction between soft and hard clay. The workers were evacuated, and the gas jet was ignited so as to burn the gas off. No sooner had the gas been burned off, however, than a 1-inch jet of water burst out at the same location, filling the shaft with several feet of water. The workers plugged the leak, only to have it break out in a different location. All work was stopped, and it was determined that no excavation of the tunnel could begin until the shaft had been bricked up to the cylinders.[7]

Problems with incoming water would continue through early May as the workers moved forward with bricking up the shaft. By May 20, the workers had succeeded in bricking up the shaft from the bottom of the sump to the top of the lowest cylinder, which was 18 feet below the surface.[8] The shaft was walled up with masonry 12 inches thick, an operation that left an inside diameter of 8 feet. Water from an underground spring would continue to fill the sump, which had to be emptied twice daily, throughout the remainder of the work, but it no longer hindered progress. With the water and gas issues now abated, at least with regard to the land shaft, preparations were made to begin tunneling eastward. A 50-foot-long passageway was dug westward, with a chamber at the end for placement of a transit to aid in aligning the tunnel.[9] Once that task was completed, the long-anticipated day finally arrived. On May 26, 1864, the miners struck out eastward. The great lake tunnel had begun in earnest.

The day-to-day operations of the project were divided into work above ground and work below ground. The aboveground work was supervised by John P. Offermann, an employee of Dull & Gowen. There were two twelve-hour shifts of mechanics and laborers, the entire aboveground work force consisting of one superintendent, one night foreman, two engine tenders, four topmen, one day laborer, and one carpenter.

Figure 12. Working at the face of the tunnel. Source: J. M. Wing, *The Tunnels and Water System of Chicago: Under the Lake and Under the River* (Chicago: J. M. Wing & Company, 1874).

As for the belowground work, which was overseen by representatives of the Board of Public Works, it proceeded around the clock six days per week and was divided into three shifts of eight hours each. The first shift—the principal mining shift—was under the supervision of George Dewer, an experienced miner. His chief duties were to keep the excavation properly aligned and to watch for the approach of bad ground, so as to be ready with bracing, or to handle any emergencies that might arise. (As we will see later, Dewer's health failed during the progress of the work, and he was succeeded by Edward Offermann, John Offermann's son, who continued until the tunnel was completed.) The personnel for this shift consisted of four miners and four "pushers," the latter charged with pushing the loaded cars to and from the shaft so as to deliver brick, cement, or sand to the workmen and move excavated earth back to the shaft to be hauled and dumped. Only two of the miners were trained men; the others were common laborers, but they quickly learned how to use mining tools in the clay.

The excavation was carried out by two of the miners working together for ten or fifteen minutes, "with more than common vigor," and then resting.[10] They were spelled by the other miners as the pushers loaded the excavated

Figure 13. Pusher at work in the lake tunnel. Source: J. M. Wing, *The Tunnels and Water System of Chicago: Under the Lake and Under the River* (Chicago: J. M. Wing & Company, 1874).

earth into movable cars brought as near as possible to the face of the tunnel. The physical conditions in the tunnel were difficult, to say the least, since they involved working in a dark, damp, confined space where the air was fouled by smoke from the miners' lanterns, by the laborers' own sweat, and by the debris of the excavation. These difficulties were exacerbated by the psychological realities of tunneling beneath the lake. One miner acknowledged that none of them were ever "entirely without fear," and the knowledge was never far from their minds that should the tunnel collapse, there would be no escape.[11]

The second shift was supervised by Herman Kroeschell, an experienced mining engineer who was also the principal inspector of mining. This shift, in addition to continuing the excavation along the same lines as the previous shift, was the "trimming" shift, during which the interior of the excavation was made to conform as closely as possible to the exact outside shape of the masonry. Kroeschell, using plummets, ranges with sperm candles, and spirit levels, directed the setting of the patterns by which the masonry was built so as to produce the lines and levels that had been given by the engineer in charge.

Figure 14. Mason at work in the lake tunnel. Source: J. M. Wing, *The Tunnels and Water System of Chicago: Under the Lake and Under the River* (Chicago: J. M. Wing & Company, 1874).

The third shift was that of the masons, under the supervision of Edward Everett, who for years had been employed by the city as a sewer inspector. The personnel for this shift usually consisted of three masons, one mortar mixer, and four to six helpers. The water for mixing the mortar was brought from the top of the land shaft in tank cars specially made for the purpose. All the bricks used in the shaft and the tunnel were carefully inspected above ground by Michael Whelan, an employee of the board. Ellis Chesbrough, with his usual attention to detail and his insatiable curiosity, took considerable interest in the different cements used during the construction. The principal cement used was known as Clark's LaSalle, but there was also a considerable quantity of cement from Falls City, in Louisville, Kentucky, and some from Akron, a village in Erie County, New York. For the most part, cement and sand were mixed together in equal quantities, and Chesbrough noted later on, in 1869, that it would be "a matter of much interest when the tunnel [was] again entered for examination, to ascertain the relative degrees of hardness and probable durability of the different kinds of cement used."[12]

In addition to the supervisors and workmen already named, James Gowen was present almost continuously from the beginning to the end of the project. His partner, James J. Dull, was rarely at the worksite but instead handled affairs back at the company's offices in Pittsburgh. And of course Chesbrough was an ever-present figure at the site, showing up daily in the beginning and then several times a week thereafter.

When the original cost estimate for the project had been made, a wage of $1.25 per day for a common laborer was assumed, but the usual costs actually paid were $2 per day for a common laborer, $5 per day for a mason, and $3 per day for an engine man. A payroll sheet from the construction of the tunnel suggests that the majority of the workers on the project were of Irish and German descent.[13]

The rate of excavation and masonry in the tunnel varied, increasing as the miners and the masons became more accustomed to the work environment. Over the course of the project, the average daily length of masonry laid was about 13 feet, but the daily rate was far less in the early going. Later on in the project, the average daily length would sometimes reach 15 feet, but only when a pair of miners was used during the masonry shift. This practice had mixed results—the excavated earth could not be removed from the tunnel while the masons were working, but the use of a pair of masons did allow the tunnel to advance 2 feet more per day than would otherwise have been possible.

In the belowground work, great attention was paid to producing a smooth surface for the tunnel as well as for the connection of the tunnel to the land shaft, so as to ensure the optimal flow of water through the tunnel. Where masonry was concerned, the challenge of producing a smooth curved surface from the shaft to the tunnel was exceptional, but the feat was accomplished with, as Chesbrough put it, "much pains."[14]

The construction of the tunnel itself followed Chesbrough's conception. The entrance to the tunnel, at the land shaft, was 6 feet in diameter, and the tunnel itself, over a distance of 20 feet, tapered down to a diameter of 5 feet. The masonry at the entrance was made of three shells of brickwork, each 4 inches thick, with 0.5-inch cement joints between them. The rest of the tunnel proper was lined with two shells of brickwork. The cavities outside the brickwork were originally supposed to have been filled with tamped earth, but this plan proved unsatisfactory, and solid masonry was substituted for tamped earth. The upper arch of the tunnel was built on a ribbed center of boiler iron, which diminished the open space inside the tunnel by only 4.5 inches and allowed room for the cars carrying away the excavated earth to go up to the face of the excavation, which was about 10 to 20 feet ahead of the masonry.

The iron center was 30 inches long in the direction of the tunnel. An approximately 2-foot length of masonry was usually laid at any one time. At first the tunnel was excavated about 1 foot above the top of the brickwork, to give the masons room to build the upper arch, but it was soon found that they could build the arch without making the excavation any larger than the space required for the brickwork. They drove the last four or five top courses

Figure 15. Elevator transporting workers for the lake tunnel. Source: J. M. Wing, *The Tunnels and Water System of Chicago: Under the Lake and Under the River* (Chicago: J. M. Wing & Company, 1874).

of brick into well-tempered cement mortar that had been thrown into the cavity, and the driving of the bricks effectually filled up the spaces that could not otherwise have been reached by hand. The lower arch, built with the use of templates or patterns of the kind used in building sewers, was usually kept 6 feet in advance of the upper arch, to allow the cars to get closer to the miners so they could more easily load them.

As Chesbrough had expected, the excavation was generally through blue clay. The excavation did not usually require bracing, so long as it was not left to support itself more than thirty-six hours. Sometimes pockets of sand and quicksand were met, but they were small in quantity and size and presented no serious problems. The clay was mostly stiff, but sometimes sections were soft enough for a miner to run his arm into it. In those instances, a little more masonry was required; but, again, this situation did not present a serious setback. More difficulty was caused when the miners met boulders, some weighing several hundred pounds, but even in these cases the miners simply removed them with hand tools or, on occasion, small explosive charges.[15]

No sooner had the work eastward begun in earnest than the project hit a snag of the human kind. On June 1, work was temporarily suspended because of the masons' demand for more money. As of June 2, work on the tunnel had progressed to only 20 feet, but once the pay dispute was settled, it moved forward at an accelerated pace. By June 5, the work had progressed to a total of 48 feet. It was also determined that the pullers could no longer handle the arduous task of moving cars laden with earth and supplies. Accordingly, the miners set about digging stables in which to keep small mules that would be lowered into the tunnel. (The project costs for 1864 included $53.56 for "horse keeping.") At the same time, tracks were laid in the tunnel to allow the cars to be pulled more easily.[16]

In the early planning of the tunnel, the main concern had been the possibility of encountering pockets of sand or gravel. These largely failed to materialize, but another danger, this one unexpected, had first been encountered in the digging of the land shaft—pockets of flammable, explosive gas. When struck by a miner's pick, and thus exposed to the air in the tunnel, these pockets of gas would ignite and cause flash fires in the upper part of the tunnel. All that the miners could do was throw themselves facedown on the floor of the tunnel, giving themselves a chance to escape "without any greater injury than singed beards and eyelashes, and blistered faces."[17] Unfortunately, however, the first severe event of this kind left one miner badly burned and closed the tunnel down for three days. Afterward, the miners became adept at discerning the presence of gas pockets by the sound of their picks. When a pocket could be detected ahead of time, a small bore was made into it, and the gas was ignited and allowed to burn itself out.

Throughout 1864, the tunnel grew longer, advancing to 160 feet in June and to 750 feet by October. At this point, it was decided to build a pair of temporary chambers, one on each side of the tunnel, and to support them with timbers and planks. Between the two chambers a turntable was installed so that the cars could move past each other, and the chambers also allowed trains of cars to be assembled. At the site for each of the chambers, gaps were left in the masonry, to be filled in later. After the first pair of temporary chambers was built, it became standard practice to build a pair of temporary chambers every 1,000 feet, and this practice greatly expedited the work—although the chambers also introduced new difficulties, as we will see.

When the human pullers reached the tunnel's second pair of chambers, the work of pulling the cars was taken over by a team of two trained mules wearing collars with small lamps attached. Barely able to fit side by side in the tunnel, they were lowered into it by way of the shaft. Once in place, they became the "engines" of a regularly scheduled train, pulling materials to the miners and pulling the excavated earth back to the shaft. Eventually they also pulled trains of sightseers and officials anxious to get a firsthand look at the "big bore."[18]

The chambers, turntables, and mules, not to mention the increasing experience of the miners, allowed the work to continue more efficiently as the tunnel grew longer. The greatest progress achieved was 93 feet in one week.

Figure 16. Turntable in the lake tunnel. Source: J. M. Wing, *The Tunnels and Water System of Chicago: Under the Lake and Under the River* (Chicago: J. M. Wing & Company, 1874).

Figure 17. Mule train in the lake tunnel. Source: J. M. Wing, *The Tunnels and Water System of Chicago: Under the Lake and Under the River* (Chicago: J. M. Wing & Company, 1874).

Only once did the miners encounter a boulder that was too large for them to break up, and it had to be dynamited. Chesbrough later remarked, with characteristic calm, that there had been "a little nervousness as to the effect of a blast under the lake, but it caused no serious disturbance either of the ground or the masonry."[19]

Despite the benefits of the chambers, it soon became apparent that leaving gaps for them in the masonry was an unworkable practice. About three weeks after the construction of the first set of chambers, large cracks in the brickwork were found on each side of its turntable. When the second set of chambers was built, efforts were made to reinforce its turntable, but the same cracking appeared. It was finally determined that the problem had arisen because of the tendency of the clay to creep toward any gap or cavity that had been left unbricked. Accordingly, workers began bricking in the tunnel across the gaps, and the problems with cracking were solved.

As the tunnel became longer, air quality became the next serious problem. Smoke from the miners' lamps and from the heat and dirt of the excavation itself fouled the air and even made it difficult for the workers to see clearly. To address this problem, long perforated tin pipes 6 inches in diameter were run along the top of the tunnel. They led above ground to the chimney of the boiler for the steam engine used to power the elevator that hoisted men and materials up and down the land shaft. In this way, bad air was pulled out of the tunnel and was replaced with good air from the shaft opening. But this solution proved only temporary, and so when the length of the tunnel had reached about 0.5 mile, the 6-inch-diameter pipes were replaced with 8-inch-diameter pipes along with a separate engine and an Alden blower fan to extract the bad air and replace it with good. The only remaining problem was that of keeping the pipes in place, since the mules kept hitting them with their heads. On Sundays and holidays, when the tunnel was empty, the ventilation process worked nicely, and visibility was good enough that sperm candles could be seen for up to 2,000 feet.[20]

Of the many challenging aspects of the tunnel, one was particularly vexing—the problem of digging in a straight line for 2 miles under the floor of the lake, without the aid of either compasses (which were useless that far underground) or visual reference points. On land, a surveyor can fix two points and then use them as constant references. Below ground, however, another way of keeping the tunnel straight had to be found as the work advanced bit by bit. That task fell to William H. Clarke, the principal assistant city engineer.

To help the workers hold to a straight line, the tunnel itself had been designed as slightly oval in shape, 5 feet wide and 5 feet 2 inches high. The first step was taken above ground, with the mounting of an astronomical transit on a tower 166 feet west of the land shaft. The transit was used to sight a line over the center of the land shaft and outward into the lake. The same transit was used underground, in the chamber to the west of the land shaft

Figure 18. Surveying the lake tunnel. Source: J. M. Wing, *The Tunnels and Water System of Chicago: Under the Lake and Under the River* (Chicago: J. M. Wing & Company, 1874).

that had been constructed for that purpose. As the tunneling began, a line was marked on the ceiling of the tunnel, in line with that established by the transit. In the early stages of the work, sperm candles were centered and used to establish site points. Once the tunnel had been extended eastward by 280 feet and had been bricked up, a 6-inch-diameter tube was sunk from the lake surface to the center of the tunnel, and the act of plumbing up through that tube established a range from the aboveground transit that had been mounted on the tower west of the land shaft. This range allowed the eventual placement of the crib, with great accuracy, at the eastern terminus of the tunnel.

Sperm candles provided the best artificial light of the day—the term *candlepower*, the precursor of our term *lumen*, was originally defined in England by the Metropolitan Gas Act of 1860 as the light produced by a pure sperm candle of fixed weight and density—but they could be seen only up to a certain distance in the gloom and foul air of the tunnel. Once the work had progressed to the point where the candles could not readily be seen, a goniometer was used to advance the center line from point to point. In essence, this instrument employed two telescopes, which could be adjusted so as to "reverse" on a point and produce a straight line between that point, the instrument, and the "back sight." The lines thus drawn on the ceiling of the tunnel were advanced about 9 feet at a time, a tedious and unrelenting task for Clarke. Adding to his challenge was the need to ensure a constant downward grade of 2 feet per mile from the eventual inlet at the crib to the onshore pumping station. Nevertheless, the tunnel slowly but surely continued its way eastward, with the miners marking the walls to indicate every 5 feet attained.

As in any type of sustained public work, especially one carried out in such extreme circumstances, technical issues were not the only challenges to be overcome. Given the unprecedented nature of the tunnel as well as the known and unknown dangers associated with the work, it is amazing that the most serious injuries were the burns sustained by a single miner and the broken ankle of Thomas McMahon, one of the masonry inspectors. In fact, the only death associated with the project was due not to the tunnel but to murder.

The fatal trajectory had begun with a series of disagreements between two foremen, Patrick Hunt and Michael Corry, over the amount of work accomplished by their respective crews. The contractors had intervened by firing Corry, noting not only the disagreements with Hunt but also Corry's "general dissipated habits and quarrelsome disposition."[21] Corry, having been discharged, arrived at the land shaft at about half past midnight on August 23 and asked permission to descend to the tunnel to collect some personal effects. On his way to gather his things, he encountered Hunt. He told Hunt that he had been fired and that he did not want there to be any bad feelings between Hunt and himself. Hunt replied that he bore no ill will toward Corry, and apparently that was not the right thing for Hunt to have said, because the exchange of words quickly escalated to the point where Corry proposed that the two of them go above ground and fight it out. They then moved toward the land shaft, and their loud altercation attracted the attention of the foreman, Edward Offermann, and of two other employees who had been looking down into the shaft. As Hunt stepped onto the platform that the workers used as an elevator, Offermann and the two employees saw Corry take a sharpened file and stab him in the left side of the neck (according to Offermann, a lamp hanging less than 10 feet away from where the two men were fighting had allowed the attack to be witnessed).

Hunt managed to knock Corry down with an iron crowbar, and, according to the report, "would have undoubtedly killed [Corry] on the spot had he not been compelled by exhaustion occasioned by the loss of blood to relax his hold and fall down lifeless by his side."[22] Offermann and the two other employees slid down the rope and, joined by some men who had been working at the face of the excavation, managed to subdue Corry. He was then brought to the surface and turned over to the police at the North Market Station. Corry was indicted for Patrick Hunt's murder on December 3, 1864, but no disposition of the case was ever reported.

Meanwhile, Dull & Gowen were having problems of another sort. Their issues were not personal but financial. In the early summer of 1864, the contractors appealed to the city for a change in how they were to be paid. Under the original contract, they were to receive monthly advances of 75 percent of the cost of the work actually completed, and no advance payment for materials. The contractors requested instead a monthly advance of 85 percent of the cost of the work completed in addition to a payment for materials delivered and on the ground but not yet put into use. They based their request on the inflated cost of materials and labor, an increase due to a change in the gold premium since the time of the contract's signing. The Board of Public Works recommended increasing the contractors' monthly advance. The Common Council first addressed the matter in August but put it over to a special meeting, which took place on September 9.

At that meeting, the Finance Committee (Aldermen Comiskey, Kahn, and Bond) reported in favor of the board's recommendation to increase the contractors' monthly advance. Ellis Chesbrough became the focal point of the ensuing debate when he was called upon to recap the current financial situation regarding the tunnel and the crib. He acknowledged that the cost of lumber and iron had nearly doubled since the signing of the contract, and he said that he agreed with the board's recommendation to increase the contractors' monthly advance. After considerable debate, and after it had been made clear that the sureties already in place for the contract would remain secure, the Common Council approved the contractors' request. But, as we will see in chapter 6, this would not be last or most serious of the financial issues that threatened the project.

With the contractors' money problems set aside for the time being, work progressed on the tunnel. By November 19, it had reached 1,000 feet, and on November 28 the contractors formally invited the mayor and the Common Council to visit the land shaft and the tunnel three days hence, on December 1.[23] On the appointed day, a majority of the Common Council, along with Board of Public Works commissioners Rose and Gindele, arrived at the site to see the progress of the great work. Mayor Sherman was ill and did not attend, but various representatives of the city's newspapers joined the official group. The contractors were represented by James Gowen, who began by leading the visitors on a tour of the aboveground facilities, which included an array of specimens that had been brought up from the shaft and the tunnel from the bottom of the lake, where, as the *Chicago Tribune* reported, "they had lain in undisturbed repose for countless centuries" until being "brought up . . . and exhibited as specimen bricks of the great fabric of the universe."[24] Then came the big moment:

> We need not attempt to describe the sensations experienced in sinking down the shaft of seventy-six feet in depth, with nothing but a wire rope between us and destruction, and an infinitesimal patch of skylight overhead; the feeling can be realized by anyone who has dreamed of falling down a yawning chasm; to others a description

Figure 19. The great Chicago lake tunnel. Source: Pamela Bannos, "The Earliest Years: The Lake Tunnel," Shifting Grounds, http://shifting-grounds.net/pre-history/tunnel.html, accessed September 9, 2016. Photograph by John Carbutt.

> would be useless. The bottom was reached in safety, and the party stepped out into the void—a long bore, five feet in diameter, and stretching away lakeward a distance of eleven hundred and twenty feet—nearly one-quarter of a mile. A platform raised nine or ten inches from the bottom gave a good foothold, but left a rather humiliating amount of perpendicular room in which to walk. But the party set forward, bent (nearly double) on exploring the farthest recesses of the gloomy vault. 'Twas a long, wearisome creep, and long before the end was reached, the knees and back began to tire, while the elevated temperature—fifty-eight degrees—made it uncomfortably warm.[25]

One after another, the groups of visitors made the trek, and all were impressed by the character and quality of the work. Some were surprised by the relative warmth below ground, in comparison to the Chicago winter above, and by the quality of the air provided by the ventilation system. At the end of the visit, the *Chicago Tribune* concluded that the work was quite satisfactory: "Yet a little while, and we shall have pure water in Chicago, uncontaminated by the filth which is constantly poured into the river."[26]

And so, at the close of 1864, the initial hurdles had been overcome, and the tunnel was becoming a reality. Day by day, the work progressed at a rate of around 10 feet, with the three belowground crews working around the clock except on Sundays and the occasional holiday. Meanwhile, back on shore, another great endeavor was under way—the construction of the massive crib, which was to become the eastern end of the lake tunnel, and which would contain the intakes allowing fresh water to flow into the tunnel, and to the waiting city. As the new year of 1865 was rung in, little did anyone suspect the challenges that lay ahead.

Chapter 6

✦

The Two-Mile Crib

The word *crib* is generally used in civil engineering to denote a protective barrier or shield around an interior space, not unlike what is denoted by the more common usage *baby crib*. The Two-Mile Crib, however, was about as far removed from a baby crib as can be imagined. Popular accounts of the day were mesmerized by the sheer size of the structure, which was likened to the city's Court House. Nevertheless, it was quite unlike anything ever before constructed.

The digging of the great lake tunnel became the focus of the public's imagination during 1864 and 1865, but what had always worried Chesbrough was the construction and placement of the crib. The city had met its match before in the temperamental waters of the lake, and Chesbrough was rightly concerned about precisely placing a crib of such enormous size on the lake's bottom. Beyond the difficulties involved in placing the crib, there was also the severity of the forces that would be having a sustained impact on the crib once it was in place—the legendary storms of the lake, the constant wave action, and especially the crushing floes of ice that covered the lake in winter. Ships, at least, had the advantage of movement, but the massive crib, which would literally have to stand the test of time, required a structure of extraordinary strength and durability.

In December 1863, Dull & Gowen hired George R. Bramhall of Pennsylvania to work in Chicago as a carpenter and oversee the construction of the crib. Bramhall arrived in Chicago the following April and began work at a salary of $4 per day. He requested and received some pay increases during the spring and summer, and he worked on the project until November 3, when he quit over continued salary disputes and went back to Pennsylvania. But he returned to Chicago in December and would prove to be indispensable to the construction and launch of the crib as well as to the sinking of the lake shaft. Despite his contributions, however, he had a rocky relationship with Dull & Gowen for the duration of the project.

Construction of the crib had been delayed early in 1864 because the necessary lumber was not available—during the early part of the navigation season, freshets in the lumber regions had caused high waters, which made

Figure 20. Construction of the Two-Mile Crib. Source: "Two-Mile Crib," Chicagology, https://chicagology.com/prefire/prefire078, accessed September 9, 2016. Photograph by John Carbutt.

the floating of lumber untenable. Not until later in the spring did sufficient lumber become available, and so the foundation for the crib could not be laid until May. It was constructed on the north bank of the Chicago River, about 800 feet west of the lighthouse. Construction continued throughout the summer and fall of 1864, to the growing interest of the city's residents, who saw a monstrosity taking shape. Although it had originally been hoped that the crib could be launched in 1864, various delays made this impossible, and it was not completed until early 1865.

The *New York Times*, drawing from accounts provided by the *Chicago Journal*, offered the following summary of materials used in constructing the crib:

538,368 feet white-pine timber.
42,000 feet white-oak timber.
20,000 two-inch white-pine plank.
18,000 two-inch white-oak plank.
500 bales of oakum.
65 tons of iron bolts as follows:
1,700 wrought-iron 15/16-inch square bolts, 30 inches long.

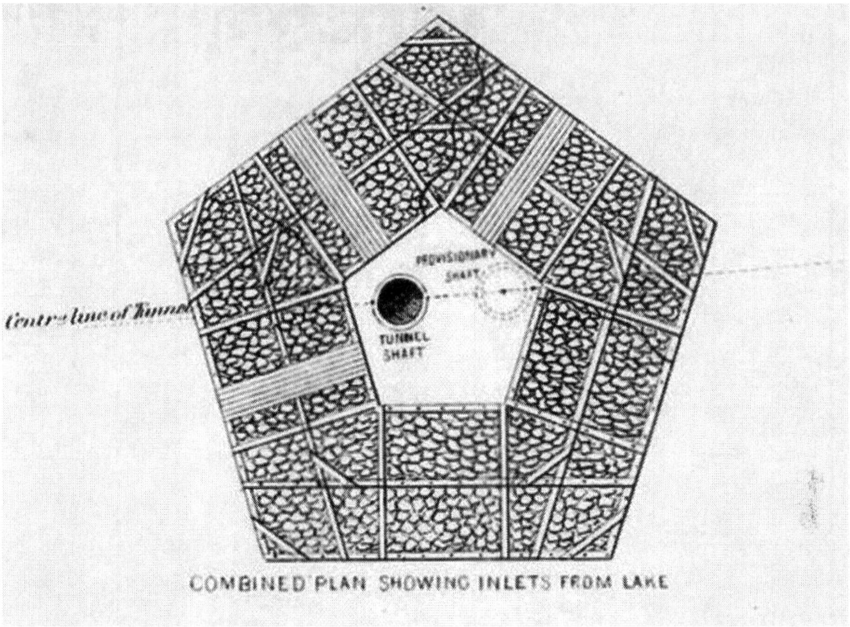

Figure 21. Detail of the wall structure and water intakes of the Two-Mile Crib. Source: City of Chicago, *Eighth Annual Report of the Board of Public Works to the Common Council of the City of Chicago for the Municipal Fiscal Year Ending March 31st, 1869* (Chicago: Jameson and Morse, 1869).

120 wrought-iron 1-inch square bolts, 30 inches long.
200 wrought-iron 1 1/2-inch round bolts, 18 feet long.
50 wrought-iron 1 1/2-inch round bolts, 18 inches long.
200 square feet 2 1/2-inch iron armor for angles.[1]

The *Chicago Tribune* placed the amount of lumber at 750,000 board feet and the weight of the iron bolts at 150 tons, bringing the entire weight of the crib to 1,800 tons even before the crib was filled with masonry and stone to anchor it to the lakebed.[2]

The construction itself was designed to maximize the crib's stability and strength. Although the original specifications had called for four cribs to aid in the construction of the tunnel, it was determined that only one, the outer terminus, would be required. The crib was pentagonal in shape, and each of its five sides was 58 feet long.[3] Originally intended to be 40 feet high, it was 6 inches higher when constructed. The basic structure consisted of three walls—the outer, the middle, and the inner—which left an interior portion for the well. This well, with walls parallel to the middle and outer walls, had sides 22 feet long, circumscribed by a 25-foot circle. In essence, the crib comprised three distinct structures, with a thickness of 25 feet between the inner face of the crib and its outer face, or breakwater.

The building of the crib, a structure belonging to both land and sea, combined construction techniques appropriate to a project that would be part ship and part fortress. The flooring and the inner and outer vertical faces of the outer, middle, and inner walls were all built of 12-inch white pine except for the upper 10 feet of the outside wall, where white oak was used to withstand ice formation. Across the angles of the outer and middle walls were brace walls 10 feet long and 12 inches thick. The middle wall on each side of the crib continued straight through to the outside wall; connecting the outer and inner walls, and passing through the middle wall, were crossties of 12-inch timber placed about 9 feet apart horizontally and 1 foot apart vertically. The ends of all the timbers, where they passed through the outer and inner walls, were dovetailed and notched into the timbers of the middle wall. Most of the joints were first sawed and then hewed to fit.

The floor timbers were laid on ground timbers placed directly under the outer, middle, and inner walls; round bolts, 1.5 inches in diameter and 36 inches long, were placed vertically, 4 feet apart, to attach ground and floor timbers to the first two courses of wall timbers above the flooring. The wall timbers were fastened to each other by bolts 1.25 inches in diameter and 34 inches long. These bolts were pointed and driven at something of a slant into 1.25-inch auger holes about 5 feet apart. Bolts nearest to each other were slanted in opposite directions to prevent their being drawn out by the buoyancy of the timber.

Three rectangular openings, each 4 feet wide and 5 feet high, were made through the crib at different depths below the surface of the lake so that water could be let in from near the bottom, the middle, or the top of the crib. These openings led to wells 4 feet square from the opening to the top of the crib, and a temporary gate was put on the inner face of the well so that the water could be cut off when necessary.

The floor and the walls of the crib were all carefully caulked, as in the shipbuilding of the day. The interior of the crib was divided into seven watertight compartments. These compartments would first be filled with stone, to anchor the crib to the lake's floor, but they were designed so that solid masonry could eventually be added to make the crib one solid mass. The outside of the crib was sheeted with 2-inch pine plank that was jointed, placed vertically, and spiked on. To strengthen the crib against the ice floes that built up on the lake in the winter, the upper 10 feet were sheeted with 3-inch white oak, and each outside corner was protected by 1-inch-thick angle irons extending 2 feet each way and fastened by 2-inch round bolts. The angle irons were let into ten pieces of 5-by-14-inch white oak that reached from the bottom to the top of the crib, fastened every 2 feet to the middle wall with 2-inch round bolts. Similar pieces of white oak were fastened by the same bolts to the inside of the middle wall.

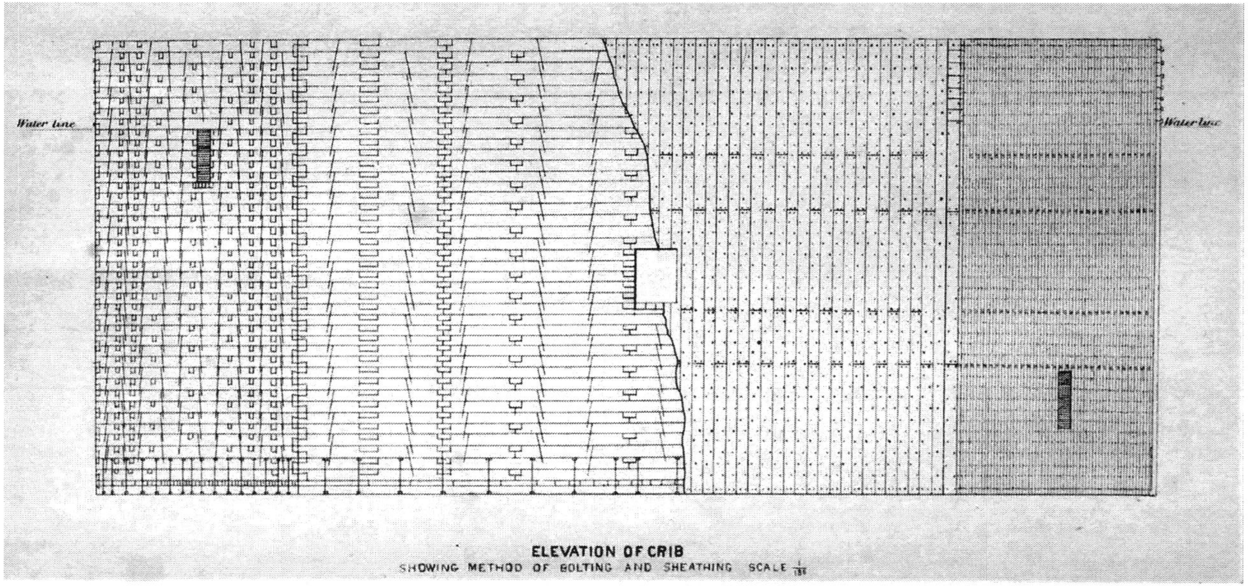

Figure 22. Elevation of the Two-Mile Crib. Source: City of Chicago, *Eighth Annual Report of the Board of Public Works to the Common Council of the City of Chicago for the Municipal Fiscal Year Ending March 31st, 1869* (Chicago: Jameson and Morse, 1869).

All in all, the crib was designed and built to withstand the almost unimaginable forces of the lake, from storms and constant wave action to crushing ice. Some thought that too much care had been given to the process of turning the massive crib into a fortress in the lake, but, as Chesbrough noted in a later commentary, "subsequent experience showed that this care was none too great."[4] Indeed, as we will see, the launch and placement of the crib led to some of the most anxious moments of the entire endeavor, and the strength and integrity of the crib's construction were put to an almost immediate test.

As we know, George Bramhall—in many ways the unsung hero of the crib's construction and launch, and of the subsequent sinking of the lake shaft—had quit his employment with Dull & Gowen in November 1864 and gone home to Pennsylvania, but in December he returned to Chicago and began work as general superintendent at the C. H. McCormick reaper factory. In January 1865, however, Dull & Gowen approached him and asked him to return to the crib project. He declined at first because of his new employment, but Dull & Gowen persisted, asking him to devise a plan for launching the crib, lowering and sinking the cylinders for the lake shaft, and generally helping to complete the construction of the crib.

Although Bramhall didn't accept the contractors' entreaties, during the winter of 1864–65 he spent his evenings, after his work at the reaper factory, devising plans for launching the crib and lowering and sinking the cylinders. He eventually submitted his plans to Dull & Gowen. The contractors and Ellis Chesbrough had already requested and received similar plans from others, but they quickly recognized Bramhall's designs as the best, and they accepted his plans. In turn, Bramhall agreed to return to the project. On April 8, 1865, he entered into his records a charge of $500 for each plan.

Bramhall returned to work for Dull & Gowen in the spring of 1865, overseeing the crib work at a salary of $6 per day. But he spent his evenings working on a plan for sinking the screw anchors that were to hold the crib in place while it was being filled with stone. These plans, too, were accepted by Dull & Gowen, and when the anchors were eventually put down, Bramhall also entered a charge for those plans into his records.

As mentioned earlier, many of the techniques used on the crib—such as those for planking, tarring, caulking, and armoring—were similar to those used in the shipbuilding of the day, and the launch of the great crib was also accomplished by methods used in the launching of large vessels. In May 1865, the Common Council had authorized the Board of Public Works to commence dredging for the laying of the "ways" to accomplish the launch. The term *ways* was short for *slipways* (a slipway being an inclined ramp or track), the means by which the crib would be slid into the river. In order for the launch to be accomplished, the entire structure of the crib had to be raised and inclined by screws (much as Chesbrough had done in raising Chicago for his sewerage plan). The slope was 1 in 12 toward the river. In preparation for the launch, seven slipways, each 140 feet long, were placed to extend 64 feet out into the river. Placing these ways was not easy; between the rocky, uneven character of the approach to the river and the accidents caused by passing vessels, it was not until mid-July that all the preparations were finally in place for the launch, which took place on July 24, 1865.

The big day began with cloudy skies and the threat of rain. James Gowen had misgivings about the weather, but he decided to press forward with the launch of the great crib, and thousands of people came out that morning to see it. Some stood on housetops, some glided over the lake on sailboats, and others watched from carriages along the banks and piers of the river.

The crib, an American flag waving from its top, was eased down onto the ways by 250 screws, and the ways were greased. Shortly after ten o'clock, all was ready. Seven tugboats steamed about the mouth of the harbor. One, the *A.B. Ward*, carried Governor Richard James Oglesby and members of his staff along with several other dignitaries in addition to members of the Common Council and representatives of the press. Waiting in the lake, aboard the tug *Continental*, was Captain A. S. Bergh, who had previously overseen the taking of borings in the lake, and who on this day was to oversee the delicate towing and placing of the crib.

A light drizzle settled over the lake's surface. Gowen, now riding atop the massive crib, gave the command, and the lines holding the crib were severed. As the crib began its slide down the ways, "cannon boomed, hundreds of steam whistles shrieked, bells rang, and thousands of throats cheered lustily."[5] It took only a minute for the crib to slide down to midriver, setting up a gentle swell as it did. A small boat piloted by two boys got perilously close to the crib, but the crib settled so quietly and gracefully on the river that their boat was not even upset.[6]

The *A. B. Ward* was the first tug to steam around the crib, and then that tug and the *Continental* each took a line to begin the slow and steady tow into the lake, where the crib would be delivered to its final destination. Despite a persistent lake swell, the initial tow was uneventful, and by early afternoon the time had come to sink the crib down to the lakebed. It was then that all the careful plans began to go awry.

In anticipation of the crib's arrival, William Clarke, principal assistant city engineer, had followed Bramhall's plan and placed five Mitchell's marine mooring screws into the lakebed. Each screw—30 inches across, 32 inches long, weighing 600 pounds, and embedded in the clay to a depth of 10 feet—had been placed opposite the intended position of each of the crib's five corners, and to each of the screws a 1.5-inch chain cable had been attached, its loose end fastened to a buoy. The plan was to attach the 1.5-inch cables to the crib while it was being sunk, thus helping the crib be properly positioned in its eventual location.

Unfortunately, things did not go as planned. A third tug, the *S. V. R. Watson,* became entangled with one of the buoys and could not get clear. Her crew had to be transferred to the *A. B. Ward,* and another boat was brought from shore to clear her propeller. Two more buoys were destroyed by other boats, which meant that only two of the original five anchoring cables were left intact. The three missing cables were replaced with regular anchors and hemp line. Finally, at two o'clock, three small grates near the bottom of the crib were opened to allow the crib to begin sinking toward the lake's floor.

During the tow, the crib's draft had been only 8 feet, but as the water flowed in, the crib's draft steadily increased to 21 feet. When the contractors felt that the crib was properly in position, they began to fill its inner compartments with loose stone, to finish sinking the crib to the lake's floor. During this process, however, the crib got out of trim, with one corner coming to rest on a sandbar on the lakebed. The contractors worked diligently to try to right the crib and get it into its correct position, but then the weather began to deteriorate.

The Board of Public Works now feared the worst. Should a storm hit while the crib was in its current vulnerable state, all could be lost. A decision was made to proceed as quickly as possible to secure the crib. As a wrecking pump moved water into or out of various compartments of the crib, trying to right it, the partitions between the crib's compartments failed, to great alarm. But the failure of the partitions proved a blessing, since it allowed the wrecking pump to be moved more readily from one compartment to another. Three tugs, using heavy tackle, were then employed to wrestle the crib into what was intended to be its final position.

With the crib now in place, the contractors immediately started filling its compartments with stone. But time had run out. A violent storm beset the crib, driving ashore the vessels loaded with stone. For three days the storm raged, every crashing wave causing a perceptible rocking of the crib. The wrecking pump was kept at work, filling the crib with water in an attempt to hold it down. As the storm battered the crib, the angle joints of the inner and middle walls began to separate. The board and the contractors could only watch as catastrophe threatened the result of their tireless efforts.

When at last the storm cleared, the angle joints were indeed found to have separated by 1 inch. The wind and the waves had moved the entire crib 13 feet, and its northwest corner was now 3.25 feet lower than its southeast corner. What Mother Nature had done, man could not undo. It was decided that the effort to try to reposition the crib would be too great, especially given the fear that another storm could easily spell the end of the entire structure. Despite the shift of the crib's position, however, it was determined that only a slight deflection of the line of the tunnel would be needed to accomplish the desired purpose. As Chesbrough later begrudgingly acknowledged, "The variation of the sides of the crib from perpendicular, though a constant eyesore, did not affect its stability."[7]

Having done the best they could, the contractors resumed filling the crib with stones and secured it to the lake's floor. This task had been accomplished by mid-August, and in the end, 9,000 tons of stone were used to fill the crib.

When first placed, the crib had been about 7 feet above the water line, but another 3 feet of height were added after the crib was filled with stone. The contractors then erected a temporary wooden structure on the crib, with a cupola on top containing a light and a fog bell. Over the quickly approaching winter, workers and inspectors would be occupied with sinking the cylinders for the lake shaft and digging the tunnel westward, and the contractors worked tirelessly to construct accommodations for them as the summer drew to a close.

The massive cast-iron cylinders that formed the lake shaft were made in Pittsburgh by James Marshall & Company, which had also made the cylinders for the land shaft. Each of the lake shaft's seven cylinders was 9 feet long and 9 feet in diameter, but, on the recommendation of the city's consultant, Erasmus Smith, their thickness was increased to 2.25 inches, by contrast with the 1.5-inch-thick cylinders used for the land shaft. The other major differences between the cylinders for the lake shaft and those for the land shaft were, first, that the lowest section of each of the lake shaft's cylinders was turned on the outside so that it would penetrate the clay more easily and, second, that the upper end of each of the lake shaft's cylinders was provided with two gateways for the introduction or exclusion of lake water. The gateways of the cylinders were 54 inches high by 32 inches wide, and they were placed so that their tops were below the lowest known lake level. Smith had put considerable effort into designing the gateways, and he provided each cylinder with a sliding gate, placed on the outside, which was raised by a screw at the cylinder's top. He also provided for the option of forming a chamber on each side of the gateway, to enable the possibility of a temporary sliding gate if it became necessary to repair the original gateways.

Other than the launch and placement of the crib, the greatest challenge at the lake end of the tunnel had to do with finding a way to secure, lower, and sink the cylinders forming the lake shaft. By contrast with the situation on land, space for machinery and a workforce was extremely limited out in the lake. Once again, a solution was provided by George Bramhall, who devised an ingenious apparatus for handling the cylinders in the confined space. So innovative was his design that on June 16, 1868, he was awarded US Patent 78,860.

Slipways had originally been installed inside the crib when it was built, but they were destroyed by the storm that struck during the crib's launch. Therefore, inclined ways were built anew inside the crib, and the lowest and next lowest of the lake shaft's seven cylinders were put together on the incline. The cylinder was held in place by chains secured to the lower end of the bottom section, and a brake over the upper end of the top section. The cylinder was lowered gradually on the incline by means of screws attached to its upper flange. Three more sections were added before the chains were removed. A false wooden bottom on the lowest section of the cylinder made it buoyant and easy to handle from overhead with block and tackle. When the cylinder was first lowered, it sank 2 or 3 feet into the clay by its own weight before its

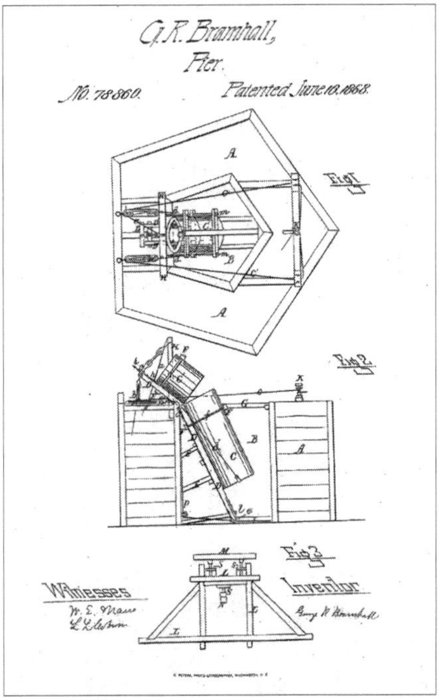

Figure 23. George Bramhall, Improved method of lowering cylinders, US Patent 78,860, issued June 16, 1868.

false wooden bottom stopped it. A hole was bored through its false bottom, and the cylinder sank several more feet by its own weight. After the sixth section was put on, the false bottom was removed entirely and the sand and clay of the lakebed were excavated. This procedure allowed the cylinder to continue to sink by its own weight. After the top section was put on, a moderate force was applied that pushed it down 23 feet below the bottom of the lake. After that, the work on the lake shaft proceeded in the same way as for the land shaft. Indeed, the work proved a bit easier because the workers encountered very little water, and no pump was required. A little leakage occurred, but the water was easily removed with buckets.

Once the lake shaft was in place, an extension tunnel was dug eastward about 50 feet, in anticipation of the tunnel's possible reach farther out into the lake at some future date. The extension tunnel was fitted with a sump and a bottom on which to place another iron cylinder, should the need arise. During the construction of the tunnel proper, the extension tunnel was also used as a turnout for the cars. In addition, a 6-inch tube was sunk from the surface, perpendicular to the extension at its outer end, and it provided a way to line the tunnel up with where it was supposed to meet the other tunnel coming from land.[8]

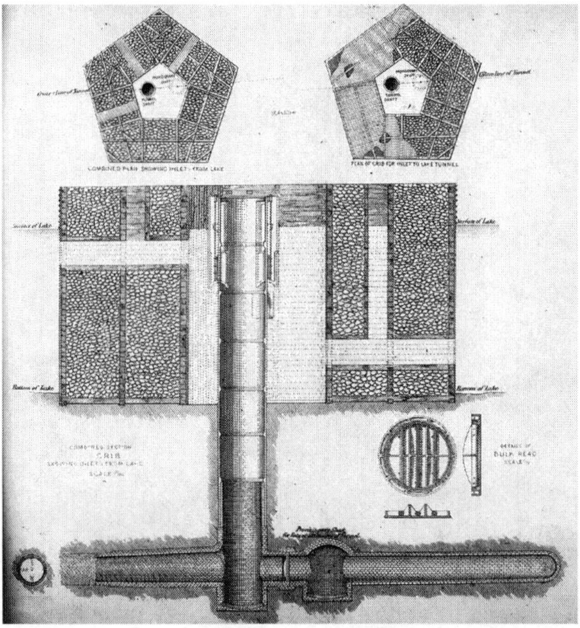

Figure 24. The crib and the lake shaft. Source: City of Chicago, *Eighth Annual Report of the Board of Public Works to the Common Council of the City of Chicago for the Municipal Fiscal Year Ending March 31st, 1869* (Chicago: Jameson and Morse, 1869).

Apart from Bramhall's innovative mechanism for placing the cylinders, the machinery and the techniques employed at the crib mirrored those that had been used on shore. The tunnel itself was to be 66 feet below the lake's surface and 34 feet below the lake's floor.[9] As on shore, a hoisting engine raised and lowered men and supplies. Given the relative shortness of the tunnel's eastern section, however, mules were not needed, and the clay brought up from down below was simply dumped into the lake.

Bramhall's contributions to the construction, launch, and placing of the crib, and to the subsequent placement of the cylinders for the lake shaft, were critical to the success of the overall project. The Board of Public Works later emphasized the impact of his contributions—but, unfortunately, the good feeling did not extend to the relationship between Bramhall and the contractors. Once the lake shaft was sunk, Dull & Gowen dismissed Bramhall. The board then stepped in and hired him directly, making him one of the inspectors of the tunneling work performed from the direction of the lake shaft.[10]

With winter now approaching, preparations were made to supply the crib with the building materials, stores, and supplies the workers would need to last out the winter months on their island in the ice. A total of forty-five to fifty men were expected to spend the winter months on the crib, working around

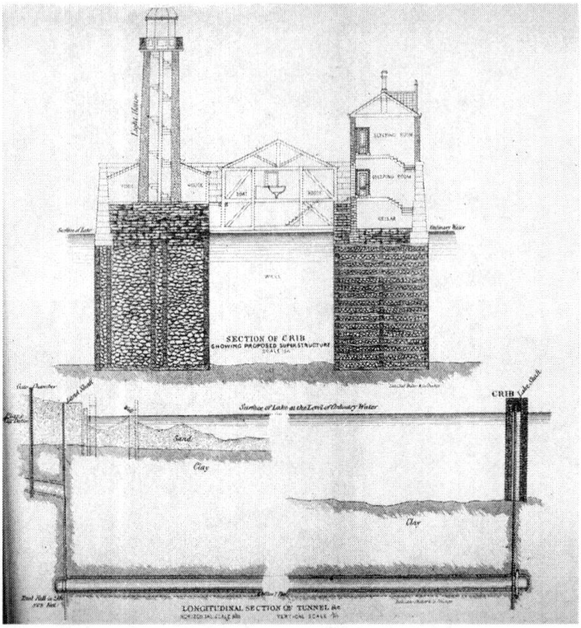

Figure 25. Sectional of the Two-Mile Crib. Source: City of Chicago, *Eighth Annual Report of the Board of Public Works to the Common Council of the City of Chicago for the Municipal Fiscal Year Ending March 31st, 1869* (Chicago: Jameson and Morse, 1869).

the clock just as their onshore counterparts had done. In anticipation of their long stay, a cook was hired and a kitchen was outfitted. Chesbrough would later note that the air was so pure 2 miles offshore that the cook complained about the voracious appetites of the men. In a rare written example of his humor, Chesbrough recounted in 1869 that the crib's reputation for providing a healthy environment must have been justified because the crib's current custodian was "quite vigorous and hearty," even though he had seemed to be "a feeble consumptive" upon assuming his post eighteen months earlier.[11]

By December, winter had asserted its grip. The men on the crib communicated their progress to the men on shore by prearranged signals and kept them apprised of the progress being made on the lake end of the tunnel. On December 22, 1865, the iron shaft was completed and the first brick was laid on the lake shaft. The westward tunneling began just before the dawn of the new year. As the ice closed in around the crib, the men had to accustom themselves to what Chesbrough called the "fearful noise" of the floes rubbing against the crib.[12] He described how, to a spectator on the crib, the crib had the appearance of "an immense plough moving through the ice."[13] Often the ice floes would mass on the south side of the crib, forming banks several hundred feet long and reaching 10 to 15 feet above the surface of the water.[14]

And so 1865 saw steady movement of the tunnel eastward from land. It had reached just shy of 1 mile out by the end of the year, with the monster crib safely placed, the lake shaft completed, and work on the westward tunnel from the crib under way. Despite the many challenges and some times of great anxiety, the project now seemed firmly in hand. But while the workers were toiling, a drama of a different sort had been playing out over 1865, one that found the Board of Public Works ready to replace Dull & Gowen and have the city take over the project.

We saw in chapter 5 that as early as the summer of 1864, Dull & Gowen had asked the city for a monthly advance of 85 percent of the cost of the work completed in addition to a payment for materials delivered and on the ground but not yet put into use. Their reason was not that the project was more difficult than they had bargained for but that the fluctuating cost of gold had significantly increased all costs associated with the project, from lumber and iron to labor costs. The Common Council granted their request in 1864, but by 1865 the contractors were back, asking for another significant increase.

In February 1865, to the surprise of many, the Common Council's Finance Committee (Aldermen Comiskey, Bond, and Kahn) recommended a 60 percent increase in the original contract price of $315,139, which amounted to a little over $189,000. This recommendation was immediately resisted by both the public and the press. The *Chicago Tribune* argued that the contractor had no reason to ask for an increase in the contract, given the substantially lesser scope of work than had originally been called for. For example, instead of four cribs and four lake shafts, the paper said, there was only one of each, and the actual tunneling had proved much easier than anticipated. Moreover, the contractors had known as early as the summer of 1864 that the price of gold was going up.[15]

The Common Council met on March 6, and the Board of Public Works strongly opposed the Finance Committee's recommendation. Ellis Chesbrough provided a cost analysis, showing that under the existing contract price the contractors would have a profit of just over $12,000. The Board of Public Works echoed the arguments made by the *Chicago Tribune*—that the scope of the work had been reduced, and that the price of gold would soon be declining. When the question was called on the Finance Committee's recommendation, it was overwhelmingly rejected by a vote of twenty-five to four.[16]

Four days later, Dull & Gowen formally petitioned the Common Council to be released from their contract with the city. This petition was taken up at the Common Council's meeting of March 20, and the Board of Public Works responded firmly, recommending that the contractor be released. The board said that the city could take over the project and either complete it on its own or put it back out for bid. Alderman Woodman moved to reconsider the Common Council's earlier rejection of the Finance Committee's recommendation to substitute an increase in the original contract price of 50 percent for the contractors' requested 60 percent. The council rejected that proposal by a much narrower vote of fourteen to twelve. Woodman tried again, this time proposing a 45 percent increase in the original contract price, but that effort failed, too.[17]

In early April, the *Chicago Tribune* strongly reiterated its opposition to an increase in the contract price, describing the contractors' request as characteristic of a "swindle" and publishing a petition, signed by a number of influential citizens, against the increase.[18] At a special meeting of the Common Council on April 26, the Board of Public Works once again stated its opposition to any increase in the contract price and beseeched the Common Council to "strengthen the hand" of the board in its efforts to enforce the contract. Alderman Woodman responded by once again asking for a 45 percent increase in the original contract price, a motion that was once again defeated, this time by a vote of fifteen to thirteen.[19]

The dispute over the contract price extended into 1866, and at the Common Council meeting of January 22 of that year, the Board of Public Works reported that Dull & Gowen were now requesting a 75 percent increase in the original contract price. The board recommended against this requested increase but did soften its view regarding whether some increase might be justified. The board noted that the contractors had lost money in their failed attempt to make bricks at the site of the land shaft, and that, given the increased costs associated with the crib, and with the stone for the crib, the contractors now stood to make a profit of only $1,279. The board therefore suggested that the Common Council consider some increase in the contract price. This proposal was referred to the newly constituted Finance Committee of Aldermen Bond, Holden, and Wicker.[20] Shortly thereafter, the *Chicago Tribune* even changed its view, agreeing with the Board of Public Works that, although a 75 percent increase in the original contract price was too much, the contractors were entitled to some modest increase in the contract price.[21] On February 12, the Finance Committee reported back to the Common Council, recommending against the 75 percent increase in the original contract price but saying that the Board of Public Works should propose an increase that would be appropriate.[22] In response, the board said at the Common Council's February 19 meeting that it agreed that some increase might be in order, but that it did not want to propose a specific increase until the work had been completed and a review of the project's finances could be made.[23]

This game of hot potato then took a dramatic turn. Despite the board's recommendation that nothing be done until after all the work had been completed, the Common Council, at the same February 19 meeting, approved a 60 percent increase in the original contract price by a vote of twenty-one to eight—an increase of $189,083. The press erupted with protests from citizens and editorial boards alike. Alderman Carter even took to the press to request that Mayor Rice veto the increase in the original contract price.[24] The matter was finally put to rest on March 5, when the mayor did just that. There would be no further discussion about an increase in the original contract price until the work had been completed, although the Common Council, with the mayor's blessing, did advance Dull & Gowen $40,000 to aid in their completion of the project. The city and the contractors, with this dispute behind them, then turned their mutual efforts to the final push to complete the great lake tunnel by the end of 1866.

Chapter 7

✦

The Wonder of America

While the politicians and the contractors fought about money, the laborers and the inspectors pushed on with their work. By mid-March of 1866, the western section of the tunnel had progressed eastward 1.25 miles from shore and the eastern section had progressed 550 feet westward from the crib, despite a string of misfortunes earlier in the month. A bad gas explosion had brought work to a halt for several days, there had been breakdowns of machinery, and some of the workers had gone on strike. But these obstacles were short-lived, and once they were resolved, work progressed steadily.[1] Tin pipes 6 inches in length had been connected to the furnace of the hoisting engine on the crib, to ventilate the eastern section of the tunnel, but, given the eastern section's relative shortness in comparison to the western section, there was no need for a bellows to supply the workers with fresh air.[2]

In addition to the problems already mentioned, Ellis Chesbrough had been obliged to deal with a very sad affair early in 1866. George Dewer, in charge of the principal mining shift from the beginning of the project, had fallen mentally ill late in 1865. He had begun to have delusional periods when he would rave about conspiracies against him, accusing even his closest friends. Chesbrough was finally forced to bring an action of insanity against him, and on January 10, 1866, the case went to trial. Dewer appeared rational during the morning session, even acknowledging his occasional derangement, but he insisted that he was cured and that his problems had been caused by the foul air in the lake tunnel. Unfortunately, he exhibited severe impairment in the afternoon session, and the jury returned a verdict of insanity. That evening, two employees of the Board of Public Works accompanied him to the train depot, from which he was to be transported to the Jacksonville Insane Asylum, but upon arriving at the station, Dewer loudly proclaimed that he was the victim of a conspiracy and was being kidnapped. He managed a temporary escape but was soon apprehended and placed on the train, accompanied by a policeman. Dewer jumped from the train as soon as it began to get under way. He was soon apprehended once again, was placed in irons, and was finally transported to the asylum.[3]

Throughout the construction of the tunnel, various visitors from near and far came to see the "big bore."[4] J. M. Wing, in *The Tunnels and Water System of Chicago* (1874), originally published in 1867 as a contemporaneous sketch of the tunnel project, recounts several of these visits; his account of a visit by a group of aldermen and other dignitaries from Boston in the late summer of 1865, shortly after work had begun on the sinking of the cylinders for the lake shaft, gives a flavor of the experience:

> Of course they must see the Tunnel, and a journey under the lake was put upon the programme, especially as the guests all requested it, saying that it was one of the principal things they had come to see.
>
> Arrived at the Water Works, the Boston men were all anxiety concerning their expected adventure, not caring to be killed outright in Chicago, as they seemed to fear they were going to be. But they were told that many illustrious gentlemen besides themselves had gone down into the Tunnel before them, and returned to the outer world alive. Lieut. Gen. Grant's staff officers and a large party of gentlemen descended into the Tunnel in July, and Sir Morton Peto's party also paid a visit to the "big bore." Where such gentlemen had been, the heroic firemen and aldermen of the "hub" could certainly go with impunity and profit; and so it was decided to undertake the subterranean and sub-aqueous journey.
>
> The first step previous to the descent was examining the hoisting machinery, and the apparatus by means of which the air in the Tunnel was kept pure. Everything was fully explained by Col. Gowan [*sic*], who took pains to carefully elucidate every point and answer every question. The machinery was beautiful in the extreme, the hoisting machine being almost noiseless in its revolution, and perfectly under the control of the engineer. It was so arranged that in case of an accident it could be stopped in an instant, midway of the descent. The immense bellows which created a current of pure air in the Tunnel was also examined, as a point of especial interest.
>
> The next thing was to get ready for the descent. Aldermanic broadcloth, fresh and glossy from the looms of Lowell, could never survive contact with the clay of Lake Michigan. Col. Gowan again drew on his wardrobe, and produced several complete suits of exploring garments. Ald. Marsh attired himself in a very jaunty suit, mostly of cast-off corduroy; and when he was dressed looked quite as much like an Indiana Hoosier as a modest alderman. Alderman Denio was less fastidious, and turned his coat inside out. Tying a pocket hand kerchief over his wise head, Alderman No. 2 was ready. The gentlemen of the Boston and Chicago Fire Departments displayed much skill in their drapery. Their quaint "get up" excited much remark, and jesting upon personal matters was brisk. All in all, a more novel-looking

half-dozen explorers never prepared to visit the crater of Vesuvius. Seen by the faint glimmer of the miners' lamps, they looked like conspirators—and yet half of them were citizens of Boston!

"Hoist away," cried the man in charge of the elevator at the shaft, and the next moment Aldermen Marsh and Denio, and a miner with a little lamp upon his cap, were rapidly descending the shaft. Those left on *terra firma* gathered around the opening in the earth and gazed after the glimmering light as it slowly went down into the bowels of mother earth; for the lake roared and tossed half a rod away, and those accompanying the taper were going beneath its waves and foam. Forsaking the sublime for the ridiculous, a wicked fellow above remarked that it was an awful lowering of aldermanic dignity, which provoked a laugh that rang out clear upon the night air. Luckily for his head, the members of the Boston city government were out of hearing.

Depositing its passengers at the bottom of the shaft, the elevator was drawn up, and the lesser lights of the party began the descent. Once upon the bottom of the shaft, which is eighty-nine feet in depth, the Tunnel proper was seen branching off under the lake like a long tube. In the little chamber where the expectant party were standing, was noticed a train of three small cars, which had been prepared to convey them out under the lake.

A diminutive mule was attached to this train, and the gentlemen seated themselves like Turks upon their curled-up legs in the bottom of the cars. One man in each car was given a small lamp, and one was hung upon the mule's collar. The driver's whip was applied vigorously, and away went the train, aldermen and all, under Lake Michigan. The Tunnel is five feet in diameter, a perfect cylinder, and just large enough to admit of the passage of the cars. Whoever raised his head above the sideboards would be sure to lose his hat, and a portion of his scalp.

Along the Tunnel ran a pipe something like a stove pipe, through which the bad air was extracted from the extreme end. As the work progressed this pipe was extended. The only caution necessary on this railroad excursion under the lake was to keep one's head clear of the pipe, which all succeeded in doing. Every thousand feet the travelers arrived at a chamber, where the miners mixed their cement and mortar, and where cars were turned round on turn-tables. As you went down into the bore, or further out, the number of feet the work had progressed under the lake was marked upon the masonry in plain figures.

Much joking was enjoyed as this singular train of cars passed further and further under the lake. No one could raise his head beyond a certain height, and could scarcely move. The mule trotted along briskly, and the current of air created by the motion was quite strong. The gentlemen from the "hub" were dumbfounded by the singularity of their situation. What if water should find its way into the bore from the foaming lake above? In such a case the descendants of Plymouth Rock would never return to the harbor where the Mayflower dropped anchor. It was horrible to think, of the bereaved Bostonians. For the nonce, all seemed to obey the mandate written over the entrance to Dante's inferno "All hope abandon ye who enter here."

The cars could go no further than the last chamber, and the party alighted. Here they found men engaged in mixing mortar and cement, by the light of little lamps hung upon sticks stuck in the clay. Two hundred feet yet remained to be traversed before the extreme end could be reached. The miner who conducted the party led the way with a lamp, and several gentlemen followed on foot. Men of ordinary stature were obliged to stoop almost double, so the last part of the journey was where the romance came in. At length the extreme end was reached—nearly a mile from the shaft.

Here all was busy labor. The sharp click of the trowel, and the dull sound of the miner's pick as it was buried into the compact clay, met the ear. The smoke from a dozen fluid lamps floated off into the long, tube-like bore, casting weird shadows over the scene. Clambering over a pile of clay awaiting transportation to the shaft by the next train, the party reached the miners. Here two stalwart men, with breasts bared and brawny muscles, were striking herculean blows against the earth before them, which yielded reluctantly, and crumbled. The air here was extremely bad, caused by the number of persons breathing it and the smoke from the lamps. The miners looked like grim specters as they plied pick and spade and trowel. For a few moments the visitors gazed awe-struck upon the inhabitants of this subterranean world. One miner said that he had been there every day for nearly two years; and so his sallow countenance and sunken, almost unearthly face indicated. . . .

The Boston men evinced a desire to lay a brick in the great Chicago Lake Tunnel, thereby immortalizing themselves, and, more particularly, that they might tell of what they had done. So an obliging miner arranged the cement, and the aldermen each deposited a brick, leaving a greenback in the itching palm of the accommodating but greedy workman. Bits of the clay were carefully picked up and pocketed as *souvenirs*, and the tired party of explorers retraced their steps. "Good evening, gentlemen," said each of the miners as the party left, showing that good manners may exist under the ground when they do not above. Returning to the first chamber, the party re-embarked on the cars, and were soon at the shaft. The elevator placed them upon *terra firma* in a few moments afterward, all safe

Figure 26. The tunnel under the lake. Source: J. M. Wing, *The Tunnels and Water System of Chicago: Under the Lake and Under the River* (Chicago: J. M. Wing & Company, 1874).

> and sound. On examining watches, it was found they had spent an hour and a half in the Tunnel.[5]

Throughout the spring and summer of 1866, the crews of workers in the eastern and western sections steadily inched their way toward each other. In early May, a shaft was sunk on the outside of the crib, 57 feet from the center of the lake shaft, and a plumb line was dropped down it so that the engineers could review the alignment. It was found that a slight deflection was needed in the tunnel from the crib because of the difficulties that had been encountered with the crib's placement. Still, this development was not viewed with alarm, since the overall alignment was very good.[6]

Newspaper accounts of the work took on the character of a play-by-play as they updated progress on the tunnel every week or so and counted down to the day when the two sections of the tunnel would finally meet. Here, according to contemporaneous reports published in the *Chicago Tribune*, are the dates by which specific hundreds and thousands of feet had been dug for the two sections of the tunnel:

April 1: 5,802 east, 645 west
April 22: 5,955 east, 840 west
April 30: 6,038 east, 915 west
May 26: 6,340 east, 1,090 west
June 2: total in excess of 7,500
June 26: 6,628 east, no report west
July 14: 6,834 east, 1,451 west

July 30: 7,025 east, 1,600 west
August 24: 7,266 east, 1,760 west
September 17: 7,561 east, 1,926 west
September 24: 7,617 east, 2,025 west

Work was stopped on the tunnel from the crib when it reached 2,290 feet. Thereafter, the paper took a different approach, simply counting down the number of feet separating the two tunnels:

October 8: 640
October 12: 540
October 23: 366
October 30: 280
November 13: 157
November 17: 99
November 25: 2

On October 7, with 640 feet remaining between the two tunnels, the engineers decided to conduct an experiment on the conductivity of the clay with regard to sound. They found that a pick hitting clay in one section of the tunnel could not be heard in the other section, whereas a pick hitting stone or metal that had been placed in the clay within one section of the tunnel could be heard in the other section. They also tried firing a pistol in one section of the tunnel, but the discharge could not be heard in the other section.[7] This experiment constituted the last in a series of observations made on sound during the construction of the tunnel. The first observation had been made when the tunnel from shore had progressed only 100 yards from the land shaft. It was found that waves hitting the breakwater of the inlet basin could be heard by workers in the tunnel through 60 feet of earth. Later, when the tunnel was 0.5 mile from shore, workers in the tunnel could clearly hear the propellers of tugs and other ships.[8]

On Saturday morning, November 24, 1866, a flag flew over the cupola of Chicago's Court House. This quiet salute was the only public acknowledgment that the two sections of the lake tunnel had finally been connected, if only through borings. Before the final 2 feet of clay were removed, several borings were made to determine the alignment of the two sections. The alignment was found to be off by only 9.5 inches horizontally and an astounding 1 inch vertically, with the western section ever so slightly higher. The difference between the overall length of the tunnel as estimated by triangulation and its actual measured length was less than 3 inches, with the tunnel being 2 miles and 7 feet long. With these measurements in hand, arrangements were made to complete the excavation.[9]

When news spread of the tunnel's near completion, there was considerable jubilation. Colonel Wood of Wood's Museum, a large institution on Randolph

Street that presented oddities and specimens from around the world, offered $100 for the first gallon of water to come through the tunnel. He would even provide the jar, he said, which he would then display in his museum for all to admire. Meanwhile, upwards of five hundred applications were received by the Board of Public Works from individuals wanting to see the "great bore."[10] Alderman Cox, perhaps trying to capitalize on this broad interest, would later suggest a plan whereby the city would charge $1 per person to tour the tunnel, the money to be used to beautify the grounds of the new waterworks facilities once they were completed.[11]

The Chicago Tribune, after having taken almost every conceivable position on the lake tunnel, returned to its earlier stance of support and admiration, especially for Chesbrough, now that the tunnel had been successfully excavated. The lake tunnel, the paper said, "is one of those works so new and strange in its plan and construction, and shows such proof of boldness and skill in the engineer who devised and has at last carried it through to completion, that it has attracted the attention and wonder of men everywhere." The paper went on to credit Chesbrough with "this extraordinary achievement of engineering, skill and sagacity" and stated that it was only because of his "unwearied fidelity, and painstaking and unremitting care and labor," that the project had been completed. Scarcely had any project "excited more admiration and wonder for its boldness and originality," said the paper, and its completion had given Chicago "a reputation for being able not only to boldly conceive, but to successfully complete the most marvelous and stupendous works." The paper lauded Chesbrough not only for his vision and engineering skill but also for the "rare fidelity and great success with which he discharges all his official duties, [and] for the entire and unassuming modesty with which he leaves to the public the recognition of his services."[12]

On the afternoon of Friday, November 30, Ellis Chesbrough and Assistant City Engineer William Clarke descended the land shaft along with a number of miners while James Dull, James Gowen, George Bramhall, and Superintendent John Offermann descended the lake shaft. Each party made its way to the face of its respective tunnel, and at twenty minutes to four o'clock the miners quickly set about excavating the last 2 feet of clay. Offermann was the first to step across the "divide" from the western section to the eastern section. There were handshakes and congratulations all around, and then the two parties returned to the land shaft, the group that had descended at the crib becoming the first to traverse the entire tunnel. With the completion of the excavation, fresh air now flowed through the tunnel, with no need of bellows or ventilation. The men arrived back at the surface to more celebration. The great Chicago lake tunnel was a reality.[13]

Although the excavation was complete, there was still much work to be done on the tunnel. The press had already been publishing complaints to the effect that the city would have to wait several months longer before the tunnel could supply water, and that the Board of Public Works had failed by not constructing the necessary pumps and wells of the waterworks that would connect to the tunnel.

The board responded with an open letter to the public that was published in the *Chicago Tribune*.[14] The board explained that even after the completion of the final excavation, many more projects associated with the tunnel alone were still to be done. For example, about 100 feet of brickwork remained to be completed at the junction of the two sections of the tunnel. In addition, the many chambers and turnouts that had been built in the tunnel at 1,000-foot intervals still had to be filled and bricked over. Once that work was complete, the tunnel would have to be filled gently with water, to test for leaks or weak spots in the construction, which, if found, would have to be repaired. Finally, the entire tunnel would have to be thoroughly cleaned and washed. As the board pointed out to the restless masses waiting for fresh water, once the tunnel had been placed into service, many years would probably go by before it could again be entered. The board said that probably two to three months remained before the tunnel would be ready for use.

Although much work remained to be done, the Board of Public Works moved ahead with plans for a ceremony to celebrate the tunnel's completion. December 6 was chosen for the laying of the final stone in the tunnel, and invitations were sent to about three hundred people, including the current members of the Chicago Common Council and those who had been serving on the Common Council at the time when the tunnel project had been approved, in addition to the members of the Board of Education, judges of the various courts, the contractors, the Board of Public Works, and sundry distinguished citizens.

The plan was to have four groups. The first party would depart from the land shaft at ten o'clock in the morning and be taken first to witness the ceremonial laying of the final stone and then on to visit the crib, where the second party would be waiting. The second party would be taken westward through the tunnel. This process would then be repeated, with the third party leaving from shore at half past noon to pass eastward, and the fourth party leaving from the crib and returning to shore through the tunnel. Once the ceremonies had been completed, the first and third parties would be taken by tug back to shore. Twenty-one cars were prepared in the tunnel for the tunnel trips, and the tug *S. H. Crawford* was readied for the lake trips.

The first party included Mayor John B. Rice, the Board of Public Works, the members of the current Common Council, Ellis Chesbrough and his immediate staff, members of the press, and a few invited guests. At the appointed time, the first party descended the land shaft while the "crib parties" were taken out to the crib on the *S. H. Crawford*. Because the mayor was to be in the first of the train of cars waiting in the tunnel, he was obliged to crawl on his hands and knees to the front. The others slowly made their way into the six waiting cars, and when all was ready the little train set out, pulled by a mule at the front and pushed by two men behind. It was soon realized that those of shorter stature would have a better time of it.

Marked all along the tunnel wall was the distance in feet from the land shaft, and when the train reached the marking 8,277, it stopped, and the passengers disembarked. Below the marking was a hole ready to receive the plain marble stone that had been prepared for the occasion. The stone was pure white, 12 by 5 by 3.5 inches in size, and a supply of premixed cement was at the ready for the big moment. Meanwhile, the dignitaries in this party were being lined up, with Herman Kroeschell (one of the primary inspectors from the beginning of the project), George Bramhall, Edward Everett (the principal masonry inspector) to the east of the hole, and with Mayor Rice, Colonel Mason (an adjunct member of the Board of Public Works), and Messrs. Gindele, Letz, and Rose of the Board of Public Works to its west. Behind them were the aldermen and the remaining guests. When all was ready, Kroeschell called the assembly to order and introduced Mayor Rice, who then addressed the party:

> Members of the Board of Public Works and of the Board of Aldermen, gentlemen contractors, and fellow citizens: At the commencement of this important work, the Mayor of the city, being its chief officer, and supposed to represent the sentiments of all our citizens, was appointed to remove the first shovelful of earth, thereby introducing the work, and assuring the world that it was the wish of the citizens of Chicago that this great work should be done. Now that the great portion of it is completed, I have the great pleasure, and the honor, as being Mayor of the city, in like capacity, to put the last finishing stroke upon this work. It is intended, as I understand it, that the citizens of Chicago seal by my act, this great work with their approval.[15]

With that, the crowd let out three rousing cheers. Everett presented a trowel to the mayor, who applied cement around the stone, remarking as he did so that he had done some other things better. He then slid the stone into its final place and announced, "Now, gentlemen, in behalf of the city of Chicago, I place the last stone in this great tunnel, the wonder of America and of the world."[16] Once again the crowd let out three cheers. The time was exactly eleven o'clock.

While workers set about finally sealing the stone in place, the mayor remarked that he had placed three copper pennies under the stone, which led others of the assembly to likewise push coins into the setting cement. When the stone was secure, the mayor said, "Gentlemen, I announce to you all that the last stone in the tunnel is laid, and the work is completed." More cheers, and then someone called out, "Long live the tunnel," to which the mayor replied "Long live the people who have built it, and may it be used by many succeeding generations."[17] This, too, was met by cheers. With that, the ceremony was concluded, and the party boarded the train for the remaining ride to the crib. Left behind was the inscribed keystone that had been placed by Mayor Rice, which read simply CLOSED UP DEC. 6, 1866.

Back on shore, flags flew from the Court House and the office of the tunnel. The guns of Fort Dearborn offered a salute, while back on the crib a return rifle salute was fired by members of the Bohemian Club. The contractors, James Dull and James Gowen, made their way to the crib with the next party through the tunnel, Gowen carrying with him a stake that had been displaced by the first shovel of dirt, back on March 17, 1864. It was not until two-thirty that all the guests were assembled on the crib for the last round of speeches. President John Gindele of the Board of Public Works mounted some stairs near the center of the crib and gave a short speech, thanking the various individuals who had contributed to the project and calling the tunnel "the most beautiful and important link in the public works of this city."[18] He then introduced Mayor Rice, who called upon the gathered crowd to rejoice that "this great work, this wonderful work, novel in its conception and perfect in its execution, is completed."[19] Aldermen Holden and Clarke followed the mayor, offering additional praise on behalf of the Common Council, as did City Attorney Daniel D. Driscoll.

After the assembled dignitaries had given their speeches, a loud call went up to hear from Ellis Chesbrough. Although he resisted at first, the crowd persisted, and finally he rose to speak:

> I have but a feeble voice. It is difficult for me to make myself heard. More than that, we who have been engaged in this work as laborers, feel that other men, and not us, should do that work of making speeches which now seems to be in order. We thank you, and we thank God, with all our heart that this circumstance has brought us together today. I feel thankful to you for what you have said concerning this work. Those who have so long and anxiously watched the prosecution of this work, while they are congratulated, congratulate you on the prospect of the coming in of a healthful stream of water as the result of your enterprise. Gentlemen, it is difficult in a matter of this kind to say what credit belongs to any person, but I think you ought to give thanks to those who have taken upon themselves the responsibility of the work. It is not difficult for an engineer to suggest a plan, but it involves great responsibility to those who employ him to carry it out. Especially is this the case in this instance. All of you know how much doubt there was about the propriety of undertaking this enterprise. Some of my most fast and near friends were of the opinion this was a most foolish enterprise. I could not blame them for such opinions, for it had the look of utopian dreaming to those who had not had the opportunity of investigating the project. But as you know the work has been carried on, and now we know as well it is a success, and upon that success I congratulate you most heartily.[20]

Gowen was then called upon to speak, but deferred to his partner, Dull, who said simply, "I have very little to say to you. I have looked into the contract I made with the city, and I find nothing in it requiring me to make a speech. If it had been there, I should not have signed it."[21] After the laughter faded, George T. Stansbury, an employee of the Board of Public Works, brought the formal ceremony to an end by reading an ode written for the occasion:

At last our feet have touched the flooring
Of the work beneath the wave;
We've left our light craft at her mooring
O'er the entrance to the cave;
Over us the waves are lashing–
Music with the wind;
Over us we hear the crashing,
Of the right arm of mankind.
We are moving to the ending
Of the work so strong and grand,
While our wondrous thoughts ascending
Through the water and the sand
Seek the presence of the being
Who this mighty power gave,
That we can, with His o'erseeing,
Tread the land beneath the wave.
God, all gracious! Thou whose kindness
Guards us now from awful death;
Thou who still dost care to mind us
Couldst destroy us with a breath;
Thou it is whose will hath brought us
All this power from our birth,
That we have a belt of waters
Passing through a belt of earth.
When the tunnel's mouth we're nearing,
And much closer seems the light,
Since we hear the word so cheering,
That the end is most in sight,
Then our strains will be in meeting,
As we step upon the strand,
From beneath the waves we're greeting
Those who stand upon the land.[22]

With the formal ceremony over, the crowd was treated to food and wine from the kitchen that had previously served the workers and inspectors on the crib. The parties that had arrived at the crib by tug made their way back

Figure 27. The Two-Mile Crib, black-and-white reproduction of a color lithograph. Source: Chicago History Museum, *Chicago Illustrated*, August 1866, ICHi-63093. Jevene and Almini, lithographers.

to shore through the tunnel, while those who had come by way of the tunnel boarded the tug for the trip back to shore. Once back in the city, many of the celebrants continued their revelry into the night.[23]

The men whose "strength of arm" had constructed the lake tunnel and the crib were not forgotten in the celebrations surrounding the completion of the project. Various efforts were made to honor them, including a large dinner at the end of 1866 and a fund-raising project that distributed money to the men in early 1867.[24] In the midst of these celebrations, the contractors and the city pushed ahead with the work of finishing the tunnel. Throughout late 1866 and early 1867, the chambers and the turnouts were filled and bricked. On February 20, 1867, the two mules that had powered the underground railroad were hoisted back up into the sunlight, and shortly thereafter the iron tracks were taken up. The ventilation piping was removed, and the tunnel was cleaned and washed in anticipation of the next big test, when water would first be let into the tunnel.

While all eyes had been on the construction of the great lake tunnel and the crib, the city had also been making preparations for the pumping works that would be needed to supply the growing city with water. Throughout 1865, DeWitt C. Cregier, overseer of the pumping station, had been visiting other pumping works with an eye toward designing a new pumping engine

Figure 28. The Two-Mile Crib. Source: J. M. Wing, *The Tunnels and Water System of Chicago: Under the Lake and Under the River* (Chicago: J. M. Wing & Company, 1874).

for Chicago. He determined that the engine would need a capacity of 18 million gallons per day. He prepared a design for a condensing coupled-beam engine, which was accepted by the Board of Public Works, and on July 20, 1865, bids were received to build the new engine. The lowest bid was from George W. Quintard, owner of the Morgan Iron Works in New York City. Morgan had built the two engines, known as 53 and 57, that were already in service at the time. The total price of the new engine, with tax, was $112,350. The city authorized the contract, and by August 1866 the larger portion of the machinery, some 200 tons, was en route from New York.[25]

Past experience had demonstrated that the water needs of the city would continue to grow, and that the waterworks would need the capacity to expand. Two ideas were considered—one was simply to add onto the existing waterworks, and the other was to build a new waterworks that would be completely separate from the old one. The main concern with the first idea was the danger that the new construction might damage the old and put the city's existing water supply at risk. The sandy soil of the site was tricky at best, and there was concern that digging the new pump wells too close to the old structures might cause a catastrophic collapse. Therefore, the idea of a new, separate works was appealing; in the end, however, the added cost and longer time of construction led the Board of Public Works to opt for extending the current works.

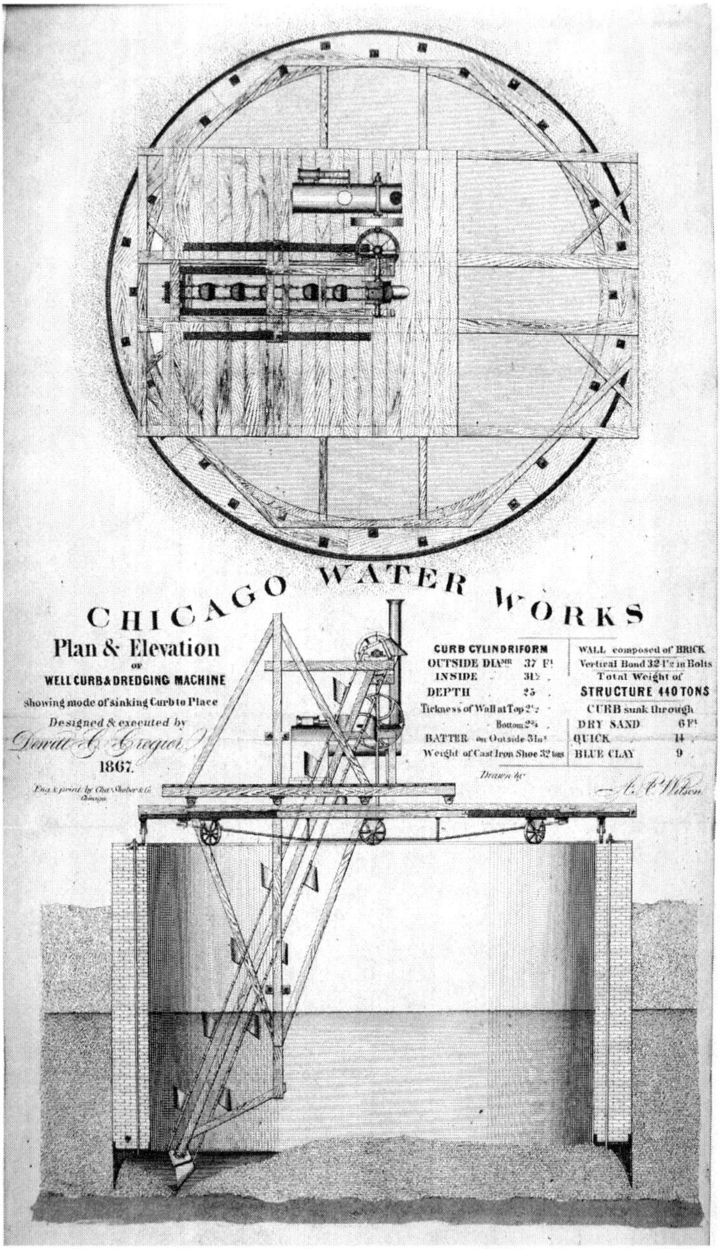

Figure 29. The Cregier dredging machine. Source: City of Chicago, *Eighth Annual Report of the Board of Public Works to the Common Council of the City of Chicago for the Municipal Fiscal Year Ending March 31st, 1869* (Chicago: Jameson and Morse, 1869).

It was a delicate business. On March 19, 1866, work began on preparing the pump well for the new waterworks, just 12 feet away from the foundation of one of the existing engines. The old works were partially demolished, and a temporary structure was put up to protect the workers. Because of the sandy soil, it was determined that the well could not simply be dug out. Instead, Cregier invented a dredging machine that would allow the well to be built on the surface and the ground then to be excavated from within the well until clay was reached. Beyond that point, the well could be sunk in the normal manner.

The curb, or wall, of the well was 37 feet in diameter and 21 feet deep. It was built from hard-burned brick and was then plastered inside and out with cement. From top to bottom, the wall was bonded with band iron every 3 feet. The base sat on a cast-iron "shoe" weighing 32 tons. Once the curb was sunk, the bottom of the new well was 20 feet below the surface of the lake and 9 feet below the bottom of the old well. It was a massive structure in itself, weighing over 500 tons.

The curb was completed and ready for sinking on June 23, 1866. Cregier's dredging machine worked beautifully, although the work was tedious, given the quicksand and water that had to be excavated from the interior of the well. Particular difficulty arose when the workers encountered an old piling from the original 1852 pump well. Once that was removed, the sinking of the well proceeded at a variable daily rate, sometimes 13.5 inches and sometimes only 3 inches. Just before reaching the hard clay, the workers encountered a large log that proved very difficult to remove, given the 13 feet of water within the well. After the log was removed, clay was finally reached, and the work accelerated, with the well sinking 3.5 feet in ten hours and, at one point, 28 continuous inches under its own weight. The sinking of the well was completed on March 28, 1867, without any damage to the well itself or to the nearby foundation of the old engine.[26]

While the new pump well was being constructed, the city also took steps to increase the overall size of the waterworks. On July 25, 1866, bids were opened for the building of a new water tower. On July 30, 1866, a contract was signed with A. Wallbaum to provide the masonry for the new tower. On September 17, 1866, another contract was signed with Wallbaum for the construction of the new engine house. To make room for the expanded waterworks, the city purchased from William Lill, for $40,000, a parcel of land just to the west of the existing waterworks. With the land secured, work began on the foundation for the new engine and engine house.[27] The first stone of the foundation was laid on October 6, 1866, and the foundation was completed on November 29, 1866.

The overall plan for the new waterworks envisioned four pumps: the old ones known as 53 and 57; the new one designed by Cregier, which would come to be known as engine 67; and an additional pump to be added at a future time. A connection was called for between the lake tunnel and these

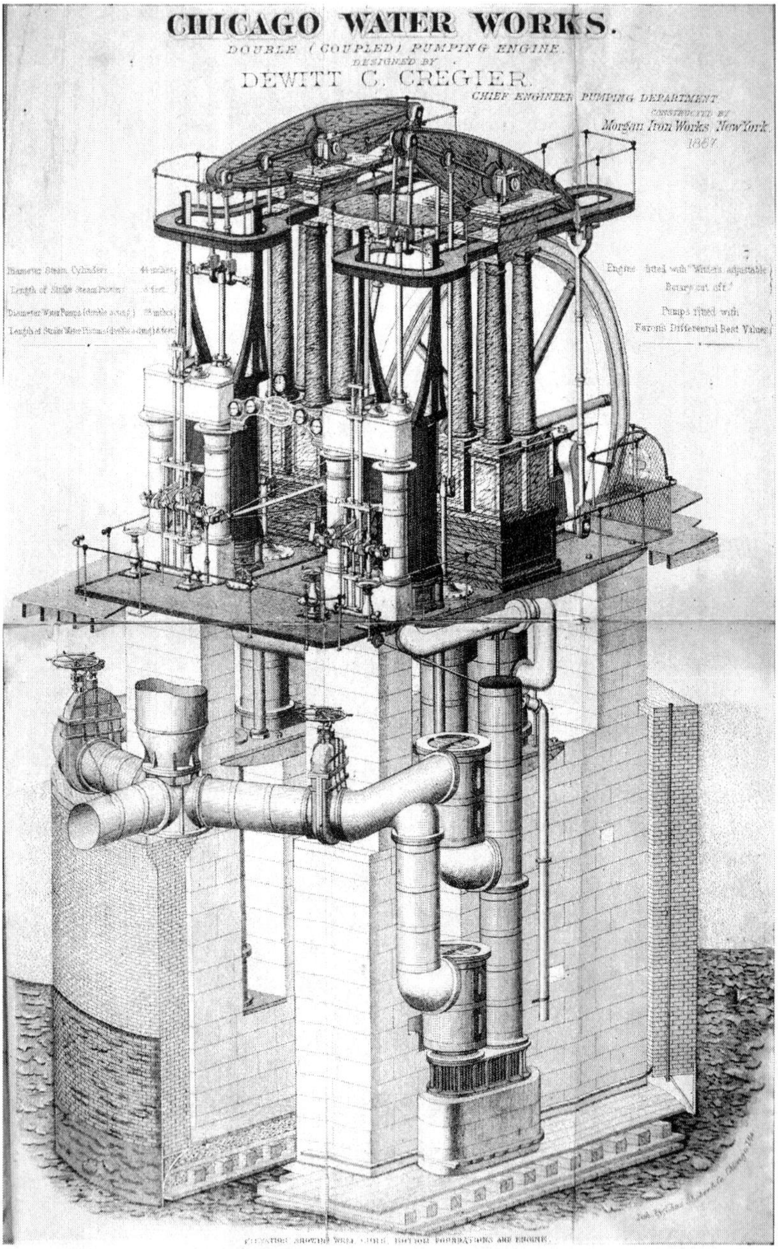

Figure 30. Engine 67. Source: City of Chicago, *Eighth Annual Report of the Board of Public Works to the Common Council of the City of Chicago for the Municipal Fiscal Year Ending March 31st, 1869* (Chicago: Jameson and Morse, 1869).

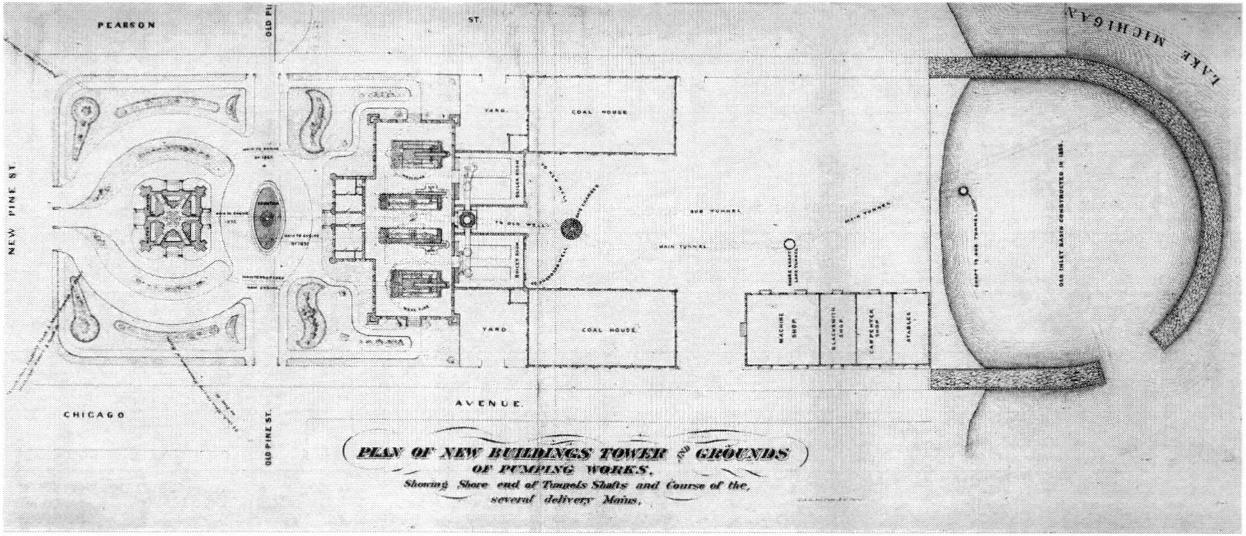

Figure 31. Plan for the new waterworks. Source: City of Chicago, *Eighth Annual Report of the Board of Public Works to the Common Council of the City of Chicago for the Municipal Fiscal Year Ending March 31st, 1869* (Chicago: Jameson and Morse, 1869).

pumps. To that end, construction began on a large gate chamber in 1866. The gate chamber would be divided into five sections—three to regulate water flow to the pumps, one to connect to the lake tunnel, and one to connect to a short tunnel leading into the old inlet basin. The latter, a backup tunnel, was intended to ensure an uninterrupted water supply for the city and was to be used in the event that the lake tunnel became unavailable for any reason.

The gate chamber had an exterior diameter of 19 feet 4 inches and an interior diameter of 16 feet. Its five compartments were separated by walls 20 inches thick. The foundations were on a bed of concrete 2 feet thick, upon which the brick footings of the exterior and dividing walls were built. At the bottom of each section were rectangular gate openings 3 feet wide and 5 feet high. The tops of the openings were 23.5 feet below the surface of the lake. The openings were fitted with tapering gates operated by rods and screws that were turned at the surface by hand wheels.[28]

The connection of the gate chamber with the lake tunnel was made via the land shaft of the tunnel by a connecting tunnel of the same size and shape as the lake tunnel. The connection with the old and new pumping wells was 4.5 feet in diameter, as were the connections to the fourth pump well that would later be built and to the backup tunnel to the old inlet basin in the lake.

So that the city could start partaking of the fresh water to be supplied by the lake tunnel while the new waterworks was being constructed, a temporary connection was made between the land shaft and the old inlet from the basin that ran to the old waterworks. Under the direction of Herman Kroeschell, a timbered horizontal drift, 4 feet square, was constructed from the land shaft to a circular brick shaft that was 4 feet in diameter and 30 feet deep. The shaft, built and sunk under Cregier's direction, had two gates just below the surface level of the lake. This shaft in turn was connected to the old inlet, and a temporary structure was erected to protect the workers and keep out intruders.[29] Under this arrangement, water would flow from the lake tunnel's land shaft to the temporary well and then through the old inlet to the existing waterworks.

When the work just described had been accomplished, the time was near for putting the lake tunnel to the test by admitting water into it. Late in 1866, stone was removed from the crib and then relaid in cement. This procedure was designed to make the structure imperishable even if the wooden timbers from which it was built were to decay. In the winter of 1866–67, the lake ice did manage to cut through some of the iron armor, all the way to the timbers, but the crib stood firm. The original iron armor was later replaced with even more substantial iron.

On March 7, 1867, it was at last time to test the integrity of the tunnel by letting in water from the lake. The tug *G. B. McClellan*, with Captain A. S. Bergh and four men aboard, set out for the crib for the purpose of raising the gates to allow water into the tunnel. They were barely 300 feet east of the lighthouse when the ice on the lake became too dense and forced them to return to shore. Undaunted, Bergh and his men gathered provisions,

Figure 32. Inspection of the tunnel. Source: *Harper's Weekly*, April 20, 1867.

descended the land shaft, and set out on foot for the 2-mile walk to the crib. At seven o'clock that evening, they raised the gates on the crib and allowed the tunnel to fill to a depth of 3 feet 2 inches, with the water flowing in at a rate of 2.5 cubic feet per second. The next day, the tug was able to make it to the crib, the gates were raised again, and this time the tunnel was completely filled and a slight column of water was admitted into the lake shaft.[30] By the morning of March 11, both the land shaft and the lake shaft were filled to the level of the lake. The water was then pumped out to the point where an inspection of the tunnel could be made. This inspection was carried out by Chesbrough himself, accompanied by three members of the press.

The *G. B. McClellan* ferried the inspection party out to the crib, where they were met by Captain Bergh. The reporters were somewhat alarmed to find that the hoisting mechanism had been removed, which meant that they had to be lowered by rope down to the tunnel. Chesbrough apparently cut quite the figure, since the *Chicago Tribune* reporter described him as being "in the guise of an ancient mariner, a broad brimmed sou'wester on his head, a rubber coat on his back, and his nether limbs protected by a pair of high boots."[31] At the bottom of the shaft waited a flat-bottomed boat. The water trickling in from the gates had partially filled the boat, and the men, sitting low in the boat, in the water that had collected there, found the voyage most uncomfortable.

They made their way by pushing along the walls of the tunnel. At the crib end, the water was halfway up the tunnel, and as they proceeded westward toward the shore, the depth of the water increased with the lowering grade. They found the tunnel to be in perfect condition, with not a brick loose. When they arrived at the keystone that had been placed by Mayor Rice, one of the reporters leaned over the side of the boat a bit too far, partially swamping it. The men bailed the boat out as best they could and continued west until they were about 1.5 miles from the crib. At that point, the water was so deep that the men were no longer able to sit upright in the boat, and so they turned around and began the return voyage, pushing eastward back to the crib. There was some alarm already growing among the reporters because only two of the lanterns that they had brought with them were still lit.

As the men drew nearer to the crib, their little vessel began to take on more water. The man at the bow suddenly sounded the alarm that the bow was going under, and when the other men attempted to bail the boat out, it capsized, sending all aboard into the frigid water. Accounts differ as to how far away from the crib the men were when this happened; the *Chicago Tribune* reporter said it was 1 mile, whereas Chesbrough estimated that the crib was 600 feet ahead. Regardless of their actual distance from the crib, this was a harrowing and dangerous situation. The men were in complete darkness, their two remaining lamps having been extinguished, and they were waist-deep in water, with their boat sunk beneath them. After considerable effort, they were able to retrieve the boat, although they were unable to make it usable, and they began to slog their way eastward, pulling the boat behind them. They called out into the darkness for help, but to no avail. The reporter described the cold as feeling like a "thousand needles."[32] At last a light glimmered in the distance—Captain Bergh, concerned at the length of time the men had been gone, had come looking for them.

The danger was not yet past. The men, doubtless suffering from hypothermia, still had to make the ascent of 68 feet by way of a ladder to the top of the crib. Chesbrough went first, and when he was halfway up, according to the *Chicago Tribune* reporter, he "emptied a gallon of water from each boot on our heads."[33] A makeshift platform had been lowered 15 feet down the lake shaft to assist the beleaguered men, and one by one they reached the platform and from there were hoisted the rest of the way up by ropes and waiting hands. Chesbrough would later recount that if the accident had occurred farther away from the crib, "it would have proved very serious if not fatal to most of the party, as the water was too cold to be endured long."[34]

With the successful inspection of the tunnel, water was let in once again on March 24, and at about four o'clock in the afternoon the mouth of the old inlet was cut off from the lake. The pumps, which were left running, pulled the water in the temporary inlet down about 1 foot, and then fresh water bubbled up in the temporary well. The great Chicago lake tunnel was ready to supply water to a thirsty city.[35]

The city had been waiting, and not always patiently, for the day when the tunnel would supply the long-promised fresh water to it citizens. If the celebration of the tunnel's completion had been more or less a private affair for dignitaries, the official opening of the tunnel, along with the laying of the cornerstone for the new waterworks, was to be a citywide gala.

The *Chicago Tribune* announced that the first flow of water through the tunnel to the city would take place on Monday, March 25, 1867, and would be accompanied by the laying of the cornerstone for the new waterworks. The paper greatly anticipated the event, noting not only that the supply of fresh water would benefit the quality of life of the city's residents but also that it would "abolish one extensive excuse for tippling."[36] The ceremonial laying of the cornerstone and the opening of the water flow were to be preceded by a procession comprising Masonic as well as civic participants. The official program for the ceremony read as follows:

> The following official order, from the Chief Marshal of the procession, contains a summary of the proceedings which will be carried out to-day on the occasion of the formal inauguration of the great Lake Tunnel, and the laying of the cornerstone of the new Waterworks on the lake shore at the foot of Chicago avenue in the North Division:
> PROGRAMME OF ARRANGEMENTS for laying the corner stone of the new buildings for the Chicago Waterworks, March 25, A.D. 1867 (A.L. 5867).
> The procession will be formed in the following order:
> Platoon of Police.
> Light Guard Band.
> Apollo Commandery Knights' Templars.
> Subordinate Lodges of F. & A. M. in the following order:
> Tylers, with drawn swords.
> Stewards.
> Master Masons.
> Secretary and Treasurer.
> Wardens.
> Masters.
> M. W. Grand Lodge F. & A. M., State of Illinois.
> Grand Tyler with Drawn Sword.
> Grand Stewards.
> Principal Architect, with Square, Level and Plumb.
> Grand Chaplains.
> Past Grand Officers.
> Junior Grand Warden, with Silver Vessel of Oil.
> Senior Grand Warden, with Silver Vessel of Wine.
> Deputy Grand Master, with Silver Vessel of Corn.

Master of the Oldest Lodge, carrying Book of Constitutions.
M. W. Grand Master, supported by Deacons and Rods.
Grand Standard Bearer.
Grand Sword Bearer, with Drawn Sword.
Great Union Band.
Chicago Commandery of Knights Templars.
Ellsworth Zouaves as Guard of Honor to Civil Division.
His Honor, the Mayor, as Orator of the Day.
Commissioners of the Board of Public Works.
Secretary of the Board.
City Engineer and First Assistant.
City Engineer Pumping Department.
Contractors Lake Tunnel.
Contractors of Machinery and Buildings.
Judges of the United States Court and other United States Officers.
Judiciary of the City of Chicago.
Heads of Departments of the City of
Chicago and Principal Assistants.
Members of the Common Council.
Clergy.
Members of the Press.
Ex-Mayors.
Board of Education.
Police Board.
Officers and Principal Assistants Connected
with Board of Public Works.
Ex-Water and Police Commissioners.
Band.
Quackenbush Collegiate Guard.
Chicago Fire Department.
Associations.
Citizens.[37]

The Masonic division was to assemble at nine o'clock in front of the Masonic Temple on Dearborn Street and march north on Dearborn to Lake Street, then west to Wells Street, then south on Wells to Washington Street, and then east on Washington to Clark Street, where it would be joined by the civic division. From there the procession would head north on Clark to Chicago Avenue, then to Pine Street, and then north on Pine to the site of the ceremony. At this point, the column would divide, to allow the various dignitaries to ascend to a platform that had been erected for the ceremony. The cornerstone was to be placed at the stroke of noon, and its placement would be followed by the principal speech of the day, to be delivered by Mayor Rice.[38]

Figure 33. Opening day for the water supply from the great Chicago lake tunnel, March 25, 1867. Source: Pamela Bannos, "The Earliest Years: The Lake Tunnel," Shifting Grounds, http://shifting-grounds.net/pre-history/tunnel.html, accessed September 9, 2016. Photograph by Copelin & Melander.

March 25 dawned chilly and dry, with remnants of a light snow from the previous evening. The ground underfoot was as good as could be expected in Chicago's busy streets, with the mud shallow and relatively easy to traverse. As the procession took shape and began its march to the site of the ceremony at about ten o'clock, thousands of people were lining the streets. The sun shone brightly at times, and everywhere there was a sense of jubilation "as though the Lake Tunnel had flooded the city with champagne and oysters, instead of pure water without little fish."[39] All around the city, business was postponed to allow residents to participate in the celebration.

The procession proceeded as planned until it reached Chicago Avenue, where the previously good ground gave way to mud that, "out of respect for boots and unmentionables," was found too deep to continue.[40] The procession took Erie Street instead, and even there it had to take to the sidewalks to avoid the mud. At half past eleven o'clock, the Dearborn Light Artillery fired a rifle salute, announcing the approach of the procession. The new water tower, which at the time of the ceremony reached just over 27 feet in height, was situated just to the east of Pine Street, and a platform had been erected on its east side, facing the old waterworks. The cornerstone was to be placed on the northeast corner of the new works, and between the new and old works the assembled masses had to navigate piles of brick, heaps of sand, and various sheds and buildings still being used by the workmen. The procession arrived just before noon, and the dignitaries took their places on the platform. Among them were officers of the Grand Lodge, Mayor Rice and former mayor Francis Sherman, Ellis Chesbrough, James Dull and James Gowen, several aldermen, and other prominent citizens. It was estimated that upwards of twenty thousand people were gathered at the site, and that many more watched from windows or had taken to nearby rooftops and trees to observe the historic proceedings.

DeWitt Cregier, who on this day wore two hats, as both deputy grand master of the Grand Lodge and chief engineer of the waterworks, began the ceremonies just after noon. Following a welcome by Grand Master Jerome R. Gorin and an invocation by the Reverend O. H. Tiffany, a number of items were placed in a hollow to be sealed by the cornerstone. Then the cornerstone itself was lowered and, after an extended Masonic ritual, it was cemented into place.

With the Masonic portion of the program now completed, President John Gindele of the Board of Public Works introduced Mayor Rice, who gave a lengthy speech outlining, from its beginnings in 1832, the history of the efforts to supply Chicago with fresh water, efforts culminating in the work of Ellis Chesbrough, "a man," Rice said, "whose name is heralded throughout the scientific world, and whose fame rests on a foundation so broad and deep that it will endure for ages."[41] Mayor Rice was followed by former mayor Sherman, who noted that barely a hundred people had attended the groundbreaking in March 1864 because people at that time did not have "very great confidence in the success of the work."[42] With that, there were loud calls for the man of the hour. Chesbrough, after being introduced by former mayor Sherman, spoke in his usual modest and brief manner:

> Gentlemen—I cannot make a speech here today. I thank you with all my heart for the kind manner in which you have received the very high compliment which the Mayor has (undeservedly in some respects) paid to me. I thank all the gentlemen with whom I have been associated—the Board of Public Works of this city and other gentlemen—and above all, I thank God for the success he has vouchsafed to us in this great enterprise.[43]

Once the cheers subsided, it was determined, because of the cold weather, to bring the proceedings to a close.

And so, on March 25, 1867, three years and eight days after the first spade of earth had been turned in the groundbreaking ceremony, Chicago had a fresh water supply flowing to it through the great Chicago lake tunnel. It would take some time for the new water flow to flush out the sediment and pollutants within the existing pipes, but the change in water quality was instantly appreciable and welcomed. On Thanksgiving Day 1867, a group of citizens presented Chesbrough with a testimonial, thanking him for his "scientific skill, combined with unwearied energy" in completing the lake tunnel.[44] Along with their thanks, they had wanted to present him with a gift, and they came up with the perfect one—they gave Chesbrough $11,000 in water bonds.

Chapter 8

✦

Never-Ending Demand

Ellis Chesbrough doubtless appreciated the accolades that came his way after the completion of the tunnel and the opening of the water supply. But he was likely more satisfied by the objective benefits that the tunnel and the water supply provided to his city. In the year after the new water supply became available, water consumption increased by more than 35 percent. In the year prior to the opening of the waterworks, average daily consumption had been more than 8.5 million gallons; in the year afterward, it was more than 11.5 million gallons.

Throughout 1867 and 1868, work progressed on the new water tower and waterworks buildings, all of which had been designed by the architect William W. Boyington. During 1867, the old water tower and some of the walls of the old engine room had to be demolished to make way for the new construction. Wooden partitions were erected in the interim, and this improvisation made for cold and uncomfortable working conditions for the engineers and the workmen and occasionally even exposed the engine and the pumps to frost.[1] Also in 1867, the connection between the gate valve and the land shaft was completed.

By the end of 1868, the new engine room and workshop were completed. The curb for the anticipated fourth engine had been built and sunk, and work was progressing on the south wing and the boiler room. During 1868, the remaining portions of the old water tower and waterworks buildings were also removed.

In addition, 1868 saw the completion of the separating walls of the gate valve, and connections were finalized between three pump wells and their corresponding operating engines. The supplemental shore tunnel, to be used as a backup if the lake tunnel needed to be closed for any reason, had progressed to about 160 feet, and the connecting tunnel to the south pump well had been extended as far as feasible, pending the well's completion. The temporary connection to the old lake inlet was discontinued, with Chesbrough remarking that it was "a great pity the beautiful, sparkling, bubbling well in the temporary shaft, viewed with delight by so many visitors, could not be made a perpetual feature of the water works."[2]

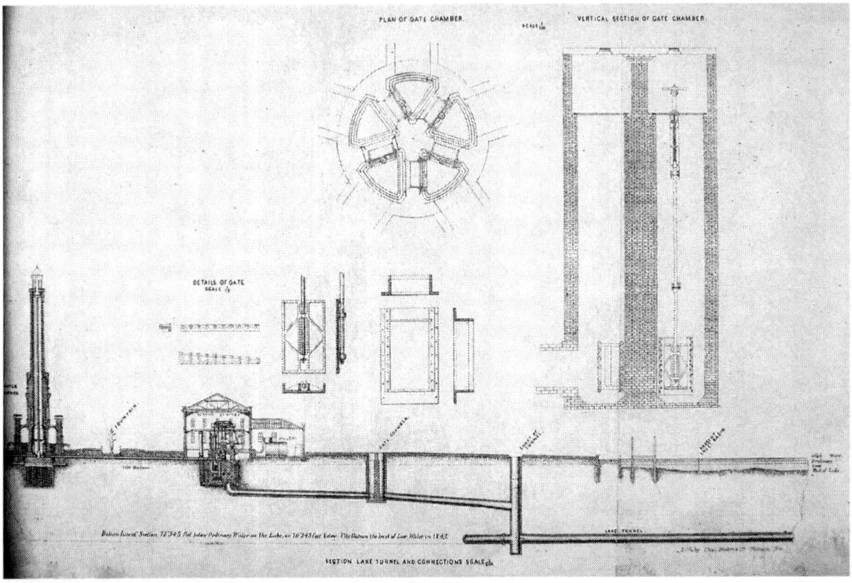

Figure 34. Waterworks sectional. City of Chicago, *Eighth Annual Report of the Board of Public Works to the Common Council of the City of Chicago for the Municipal Fiscal Year Ending March 31st, 1869* (Chicago: Jameson and Morse, 1869).

The lake tunnel itself performed perfectly, and the only incidents regarding the Two-Mile Crib were some damage caused by ice to the sheathing of the crib and some storm damage to the temporary wooden structure on top of the crib. All of this damage was easily repaired. The only functional difficulty was that the fish screens that had been installed to keep fish out of the water supply were blocked by ice at one point. A partial stoppage resulted, but the problem occurred at night and was remedied by removal of the screens in such a way that the disruption went unnoticed. The quality of the water itself was uniformly good, although it was sometimes muddied by strong easterlies or slightly fouled by an extraordinary freshet in early spring that carried waste out beyond the crib. The latter circumstance validated Chesbrough's earlier belief that an inlet merely 1 mile out into the lake would have been insufficient. Apart from these minor inconveniences, which were short-lived, the only variations in the quality of the water had to do with its differing temperatures from summer to winter.[3]

The new Cregier engine, started in July 1868, brought the cumulative pumping power of the new waterworks to 38 million gallons per day. But even with this increased capacity, both Chesbrough and DeWitt Cregier were concerned about the seemingly never-ending demand being placed on the system. In July and August of 1868, there were periods when the average daily consumption was 18 million gallons, and within those periods there were times when demand approached a rate of 20 million gallons per day. Both engineers believed that the theoretical aggregate capacity of the engines and pumps should be at least twice the maximum demand, which meant that no sooner had the Cregier engine come into service than demand was already sufficient to justify further enlargement of the system's capacity. Cregier noted that in the preceding decade, the rate of water consumption had doubled every five years. That fact, plus the increased demand caused by the improved water quality, suggested that the existing waterworks would soon meet its limits. Cregier's succinct summary of the situation was that the extension of the waterworks would be an ongoing project, one that would grow with the city, and that its completion would "occur simultaneously with that of our city."[4] Chesbrough, in a similar vein, noted the tunnel's capacity to provide more water than originally estimated, and he concluded, "It is a mere question of time, and not a very long time either, as to when a new tunnel will be needed."[5]

By October 1869, the new water tower and waterworks buildings were almost complete. They were built in a castellated Gothic style, with heavy battlemented corners, and presented a fortresslike character. The engine room was 142 feet long, 60 feet wide, and 36 feet high from floor to ceiling. It was divided into two stories, the upper floor containing drawing rooms and apartments for the engineers and the lower floor containing a reception area, offices, and the pump room itself.

The walls of the main building were 2 feet thick. The ceiling, divided into square panels, was supported by large Gothic brackets resting on heavy corbels that were built into the walls. The roof of the main building was constructed of massive timbers and covered with slate. Midway between the floor and the ceiling a balcony, protected by a Gothic iron railing, ran around the interior space and gave visitors a view of the pumping works below. The boiler rooms were in the rear of the building, and a single smokestack was placed between them. The base of the smokestack was 18 feet square up to a point 30 feet high, where it became octagonal, with a gradual taper terminating in a castellated cornice 130 feet from the ground. The exterior of the smokestack was made of stone, corresponding in style to the buildings, and the interior had a parallel circular flue built of brick. The brick flue had been built independently of the stonework, and there was an air space between the two. The entire structure rested on piles driven into the clay.[6]

The crowning jewel of the new waterworks was the water tower, a structure that became an iconic symbol for Chicago as soon as it was built. There were fluctuating levels of demand on the city's water supply, and the purpose of the water tower was to equalize hydraulic pressure in the water mains. The water tower stood 106 feet west of the main buildings, and it had been built on 168 piles capped with 12-inch oak timbers. This foundation was filled with concrete up to the surface of the water, at which point solid stone, laid in cement, continued to a point 6 feet below grade. Farther down, the gate, the pit, and the arched ways for connecting to the mains were formed on each corner. The tower itself, like all the other buildings, was of native limestone quarried in Joliet, Illinois. The base of the tower was 22 feet square, and the exterior of the shaft was octagonal.

Although contemporaneous accounts place the height of the water tower at 154 feet, the National Park Service measured the tower in 1994 and found it to be 182 feet 6 inches from the ground to the top of its spire. The tower was capped with an iron cupola that had numerous windows providing a view of the lake, the city, and the surrounding countryside. In its description of the waterworks and the tower, the National Park Service noted the influence on Boyington of the Cloth Hall in Bruges, Belgium. The tower's fancy exterior belied the underlying simplicity of the design. The first floor of the tower was essentially a grouping of four squares into a cube. Two additional cubes, similar in size, were placed vertically on this base, and on these the octagonal tower had been built. In sum, the National Park Service concluded, "Few American landmarks can project an image so graceful as this simple candelabrum format."[7]

The exterior of the tower was divided into five sections. The first section was 40 feet square, exclusive of battlements and turrets, and surrounded the base of the shaft, forming a continuous vestibule 9 feet wide on the four sides, with a grand entrance on each side. The floor and the roof of this portion were of massive stone, and the roof formed a balcony. The bottom of the interior was hexagonal, and it was here that the base piece of the standpipe (a casting, weighing 6 tons) was placed. The base had six openings, each supplied with a 30-inch gate, to which the water mains were connected. The standpipe itself was of wrought iron, 3 feet in diameter and 138 feet high. Around the standpipe was an iron spiral stairway leading to the cupola on top and lighted with alternating windows. The tower, like the engine house, was constructed wholly of stone, brick, and iron and was therefore fireproof.

At the start of 1870, the new waterworks was basically complete, although it still needed a new engine for the south pump well. Chesbrough had requested the new engine in August 1869, and the Common Council had authorized it on September 10.[8] The need for increased capacity was clear; during the summer of 1869, daily demand had sometimes reached 28 million gallons.[9] Chesbrough reported to the Board of Public Works, "The rapid increase in the consumption of water continues to be not only enormous but, considering the pumping power at present available, truly alarming."[10] Demand had already matched the combined capabilities of the old engines known as 53 and 57. In addition to the substantial amount of water that was being wasted, there was fast-growing demand for an increased supply of water to the western division of the city.

Accordingly, at the close of 1869, the Board of Public Works recommended that a second lake tunnel be built adjacent to the existing one, and that an extension tunnel be dug to the western division so as to supply a new waterworks facility. This project would be but the start of a long and continuing expansion of Chicago's waterworks in the face of never-ending demand.

Figure 35. The Chicago water tower. City of Chicago, *Eighth Annual Report of the Board of Public Works to the Common Council of the City of Chicago for the Municipal Fiscal Year Ending March 31st, 1869* (Chicago: Jameson and Morse, 1869).

Chapter 9

✦

Water and Fire

The Chicago lake tunnel was one of the most ambitious engineering feats of the nineteenth century. The actual construction of the tunnel itself proved to be easier than expected, given the uniformity of the clay along its route. But that could hardly have been known beforehand. As the *Chicago Tribune* put it, "The work on the tunnel was comparatively nothing; it was the inception and the daring which determined to carry out the idea to a successful result, which are among the wonderful."[1]

As is true of many public works projects, the magnitude of the tunnel could be fully appreciated only after its completion, and this appreciation quickly extended beyond the United States. The lake tunnel project was a highlight of the 1871 Universal Exposition in Paris, prompting William Phipps Blake, US commissioner to the exposition, to single it out for special attention in his report on the exposition. He wrote that it was "one of the most novel, successful, and economically executed engineering enterprises of the present time."[2] The Board of Public Works provided "elaborate drawings, on a large scale," for the exposition, drawings that detailed the tunnel, the crib, the waterworks buildings and tower, and the layout of the grounds.[3]

In preparing his report, Commissioner Blake relied on documents provided by the Board of Public works and on manuscript notes and explanations provided by Ellis Chesbrough. As part of his report, Blake outlined the time and materials that had gone into the building of the tunnel (see table 2). Blake also included the final costs for the project (see table 3).

After the completion of the tunnel, the Board of Public Works conducted a review of the books and statements of Dull & Gowen in connection with the contractors' request for an increase in payment for the project. The original contract price had been $315,139; the final settlement was for $380,784.60, which included $27,420 for extras on the tunnel proper and $41,225.60 for extras on the shafts, the crib, and the east and west extensions. According to the board's review, Dull & Gowen's actual cost was not more than $330,500, which meant a profit of approximately $50,000 for the contractors.[4] The crib and the lake shaft cost $117,500, the land shaft cost $12,000, and the west extension and the connection with the gate chamber cost $6,000. These items

Table 2. Time and materials for Chicago lake tunnel

ELEMENT	DAYS OCCUPIED	EXCAVATION (LINEAL FT.)	BRICKWORK (LINEAL FT.)	NUMBER OF BRICKS	NUMBER OF CEMENT BARRELS
Extension west	37	144	144	43,193	259.25
Land shaft	31	n/a	n/a	32,000	114
Lake tunnel*	953	10,567	10,567	2,393,224	9,582.25
Lake shaft	17	n/a	n/a	10,207	31
Extension east	19	65	65	29,880	106
Chambers**	35	n/a	n/a	51,250	246.5
TOTALS	1,092	10,776	10,776	2,559,754	10,339

*Lake tunnel measurement from center of land shaft to center of lake shaft.
** N = 8.

Source: Adapted from William Blake, *Civil Engineering and Public Works, Paris Universal Exposition 1867: Reports of the United States Commissioners* (Washington, DC: Government Printing Office, 1870), 29.

had not been included in the original contract, and so the cost of the tunnel itself was $195,000, which worked out to a cost of $18.45 per lineal foot, as compared to Chesbrough's original estimate of $13.54 per lineal foot. Chesbrough attributed the increase, in large measure, to the increase in labor costs over the life of the project.

From the beginning, the lake tunnel had been one element of the city's parallel efforts to manage its waste and provide a stable, clean water supply for its residents. With the completion of the lake tunnel and the waterworks, the city could point to considerable success in the latter endeavor, but it was less successful in the area of waste management. The condition of the Chicago River became worse and worse over the years, less because of sewage than because of the refuse and pollutants discharged into it by the industries along its banks. Over the years, various plans for flushing canals had been proposed as a way to cleanse the river's north and south branches, but those plans had never come to fruition. As a result, the pumping station at Bridgeport remained the only viable option for cleansing the river, and only the south branch at that.

Chesbrough had long contended that the ultimate solution to the city's waste-management problem would be the deepening of the Illinois and Michigan Canal. In 1865, the city had received permission from the state to deepen the canal enough for the Chicago River to flow into it and carry the city's waste south into the Des Plaines River. That project would take years to accomplish, however, and it was not until July 15, 1871, that the locks at

Table 3. Final costs for Chicago lake tunnel

TASK	COST ($)
Engineering and superintendence	28,744.02
Printing and advertising	375.13
Miscellaneous	6,250.03
Labor	2,096.20
Lumber	1,142.72
Piles	1,258.29
Hardware	53.85
Castings	597.55
Lake shaft gates and chambers	12,629.86
Dredging for crib	1,500.00
Tugs for board and employees	6,718.17
Discount on bonds	14,685.35
Opening celebration	979.18
Dull & Gowen, contractors	380,784.60
TOTAL	457,814.95

Source: Adapted from William Blake, *Civil Engineering and Public Works, Paris Universal Exposition 1867: Reports of the United States Commissioners* (Washington, DC: Government Printing Office, 1870), 29–30.

Bridgeport were removed and the reversal of the Chicago River's flow was accomplished. The cost to the city of deepening the canal was $3,300,883.71. Much to the city's consternation, however, in the summer and fall of 1871 the Ogden-Wentworth Ditch was built from the Des Plaines River to the west fork of the Chicago River's south branch, for the purpose of draining Mud Lake and making its former site into habitable (and marketable) land. In 1872, a heavy flood enlarged the ditch, with the result that water entered the Chicago River from the Des Plaines River during freshets, essentially nullifying the deepening work on the Illinois and Michigan Canal and significantly lessening that work's cleansing effect on the south branch of the Chicago River. The Ogden-Wentworth Ditch would remain a source of contention between the city and the ditch's owners for years to come. Given the impact of the ditch, in addition to the ever-increasing amount of sewage and waste

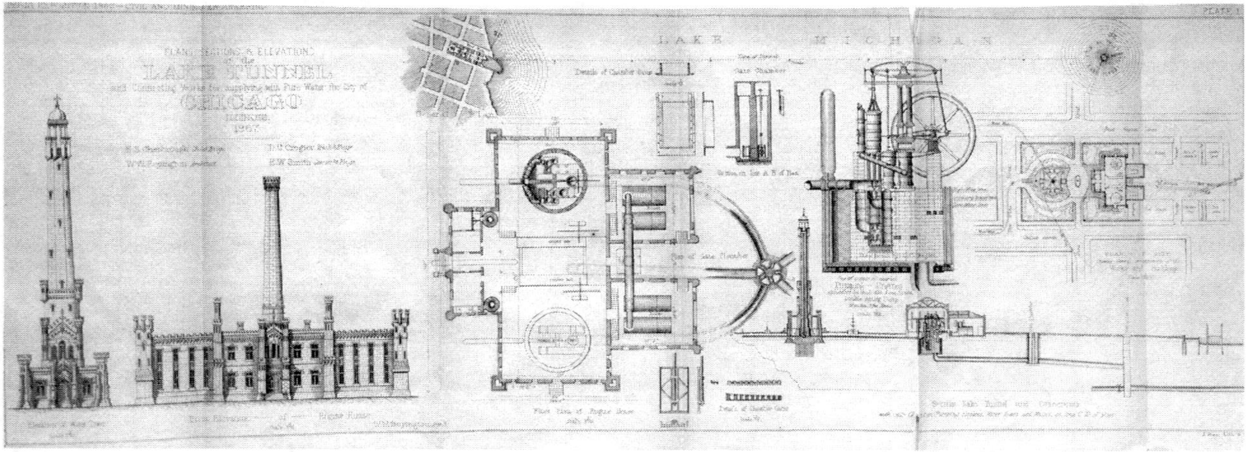
TUNNEL
CHICAGO
1867
LAKE

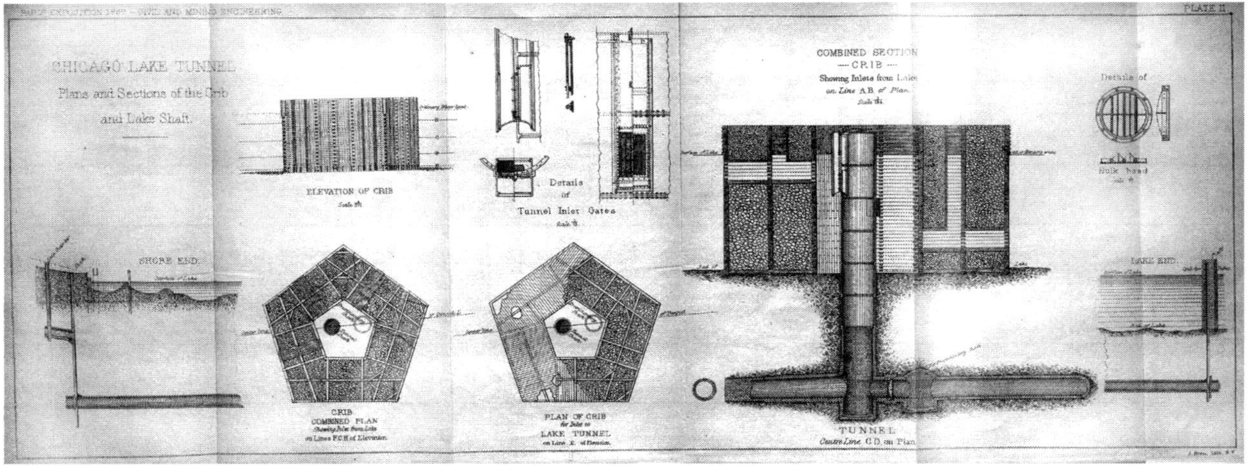

Figures 36 and 37. Chicago lake tunnel exhibits at the Paris Universal Exposition of 1867. Source: William Blake, *Civil Engineering and Public Works, Paris Universal Exposition 1867: Reports of the United States Commissioners* (Washington, DC: Government Printing Office, 1870).

Figure 38. The waterworks before the Great Chicago Fire. Source: "Chicago Water Works," Chicagology, https://chicagology.com/wp-content/themes/revolution-20/Engravings/waterworkskurz.jpg, accessed September 9, 2016.

being put into the river, it was only a matter of time until the canal's deepening alone no longer sufficed. This reality eventually led to the construction of the Sanitary and Ship Canal, at the turn of the twentieth century.

By early October 1871, in the words of DeWitt Cregier, "The machinery, the buildings and all connected with [the water] department were never in better condition."[5] The tunnel connecting the south pump well to the gate chamber had been completed, and so had the foundation and a substantial portion of the new engine for that well. The grounds were now enclosed by an iron fence, a dome had been added to the cupola of the water tower, and a "convenient and tasty drinking-fountain"[6] had been placed inside the building for the many visitors who came to see the waterworks.

* * *

On Sunday evening, October 8, 1871, at about nine o'clock, a fire broke out in a stable behind the house of Patrick and Catherine O'Leary's house on DeKoven Street, on the city's southwest side. It had been a hot, dry summer in Chicago, and that section of the city was almost entirely made up of wooden buildings. Firefighters were dispatched, but they had been given the wrong address. By the time they arrived at the source of the fire, it had already spread. The Great Chicago Fire was under way.

A strong southwest wind quickly spread the flames north and east. The fire, which eventually covered an area about 4 miles long and 0.66 mile wide, destroyed virtually everything in its path. Chesbrough's own home was destroyed. In a heartbreaking letter to his son, sent shortly after the blaze, Chesbrough wrote, "When you left us so recently we had no idea that 317 [North LaSalle Street] would be blotted out of existence. The calamity is so tremendous that we are unable to realize half its terrors." He told of loading the family's belongings onto wagons to try to save them from the fire, but, he said, "As I left the house the heat was suffocating and . . . part of our last load took fire and had to be thrown off."[7] The fire burned at a rate of about sixty-five acres per hour, and when it was finished, the property destruction was estimated at $7.5 million.[8]

The fire reached the Chicago waterworks early in the morning on October 9. Chief Engineer Cregier, who had watched the waterworks grow from a 25-horsepower engine and a pipe in the lake in 1853 to the nation's most beautiful and esteemed water facility in 1871, witnessed the inexorable advance of the fire. The grounds were bounded by open streets on three sides and by the lake behind, and so it was hoped that the stone and slate of the buildings would withstand the blaze. But a shower of sparks and embers eventually ignited the roof of the main building, and the fire quickly spread to the floors and other woodwork. Just to the south, what was now Lill's Malt House and Brewery caught fire, and the wind drove the flames from those buildings toward the waterworks. The fire became so intense that the workers who had been trying to fight the fire had to abandon their posts, and before long, as Cregier later recounted, the entire neighborhood was a "sea of fire."[9]

During the fire, the ropes that suspended the levers of the safety valves were burned away, and the valves closed, creating a very dangerous situation in which the boilers might explode. Fortunately, the fire was so intense that it melted the main steam pipe, and the steam was able to escape. By three o'clock in the morning on October 9, the fire was over, and the waterworks was "an utter wreck," with "nothing but the naked walls of the building and the broken and blackened skeletons of three engines . . . left to mark the spot from whence only a few hours before flowed millions of gallons of pure water for the comfort and convenience of our citizens."[10]

The damage was extensive, as were the associated costs to the pumping station and the distribution system. The walls of the engine house withstood the blaze, but the roof, the floors, and other portions were all destroyed. Damages to the building and the machinery were estimated at $75,000. The machine shop, a substantial brick structure two stories high that contained valuable machinery, a blacksmith and carpenter shop, and stables, was almost a complete loss, with damages set at $25,500. The north and south reservoirs were both damaged, at a loss of $20,000, and the south reservoir was never used again. In addition, more than fifteen thousand water

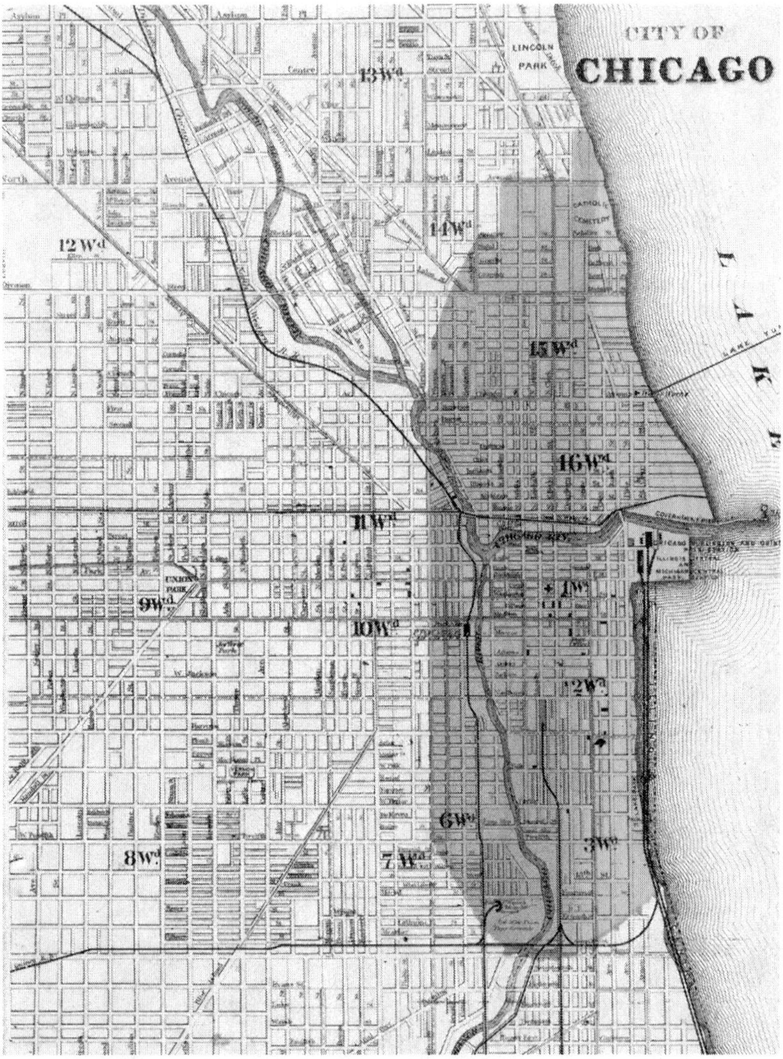

Figure 39. Map of the burned district of Chicago, 1871. Source: Chicagology, https://chicagology.com/wp-content/themes/revolution-20/chicagoimages/satoct71871.jpg, accessed September 9, 2016.

Figure 40. The waterworks after the Great Chicago Fire. Source: Chicago History Museum, ICHi-13918. From an 1871 stereograph, photographer unknown.

service pipes were either melted or damaged enough to allow serious leaks. The expense of finding and stopping the leaks amounted to $15,000, and that work was made even more difficult because the books showing the locations of the water pipes had been destroyed. Almost all the fire hydrants within the burned district had been damaged, with the cost of repairs estimated at $10,000. Similarly, about 370 water meters had been damaged, at a cost of $6,000. In addition to the damaged infrastructure, an estimated $97,410 worth of water was wasted for the six months ending on April 1, 1872, because of open and damaged pipes and hydrants.[11] Only the water tower, built almost wholly of brick and stone, managed to escape serious damage.

The fire was still smoldering when the water department, with its usual tenacity and ingenuity, began the critical work of rebuilding. A temporary water supply was taken from the river, and a boiler from a locomotive was used to provide the steam that ran the pumps. On October 10, the city began clearing the ruins of the waterworks. When workers were finally able to return safely to the building, they found that not just the engines but also the foundations were seriously damaged. Many of the engine parts were broken beyond repair, and patterns had to be made before new castings could be obtained. Almost all the joints of the steam and water pipes had been destroyed along with nearly all the lead joints connecting the water mains within the building. Fortunately, however, the boilers were only slightly damaged.

In all, more than 175 feet of 17-inch copper steam pipe had been entirely destroyed, much of it melted into a shapeless mass. Several of the stone piers forming the engine foundation were so badly burned that they had taken on the shape of an hourglass and needed to be almost completely rebuilt. During the three days after the fire, between two hundred and three hundred men were at work on the premises. In addition to laborers, the workforce included engineers, machinists, boilermakers, coppersmiths, masons, carpenters, and riggers. They worked night and day to restore the city's water supply, and, at twenty minutes past eight o'clock on the evening of October 17, the north double engines were restarted.[12]

During 1872, the city labored tirelessly to restore the waterworks. It widened Chicago Avenue by 100 feet to provide more space around the existing pumping station and diminish the risk of damage from future fires. But even as Chesbrough and his team grappled with repairing the existing works, they were focusing on the future. On May 4, 1872, the city signed a contract with Steel & McMahon for the construction of the second lake tunnel, to be dug and connected to the Two-Mile Crib. The city also purchased land for a new pumping works that would serve the southwestern portion of Chicago.

By June 1873, the waterworks buildings had been almost completely renovated, and the city was even able to begin grading and landscaping the grounds. The new engine for the south pump well had gone into operation in November 1872, and on May 30, 1873, Chesbrough was able to report to the Common Council, "The pumping power at command now is sufficient for all the ordinary wants of the city, and the ability to deliver water to the takers is decidedly greater than ever before."[13]

Although the Great Chicago Fire of 1871 severely disrupted the city's water supply system, the source of that supply, from the Two-Mile Crib through the tunnel, had remained secure. Through the exceptional efforts of the city, coupled with an outpouring of benevolence and assistance from across the country, the machinery of the waterworks and the water distribution system were made better than they had been before the fire. The city's steady growth, and the concomitant increase in sewage and industrial pollution, continue to present city officials with the dilemma of supplying Chicago's residents with clean water while managing the city's waste. But as the waterworks and the distribution system grew apace with the population, for decades to come the Chicago lake tunnel—"the wonder of America and of the world," as Mayor John B. Rice had called it—endured as an integral element of the ongoing efforts to provide an uninterrupted flow of clean water to the city beside the lake.

EPILOGUE

Ellis Chesbrough was fifty-eight years old at the time of the fire. He had seen the plans for his two enduring contributions to Chicago—the raising of the city to allow for a comprehensive sewerage system, and the building of the Chicago lake tunnel to supply the city with fresh water—become reality. He had also been largely responsible for keeping up the call to deepen the Illinois and Michigan Canal, an operation that ultimately led to the reversal of the Chicago River. His place as one of the great engineers of the nineteenth century was already secured, and his temperament and modesty had brought him widespread admiration and affection from the city's residents. He was not one to rest on past accomplishments, however, and he continued his service to the city until 1879.

Ever the careful man, Chesbrough had based his design for the waterworks on his estimate of the city's needs, taking account of the fact that all previous estimates of the city's growth had fallen short. Moreover, water consumption had always outpaced the city's growth. From 1858 to 1871, for example, while the population was increasing by a yearly average of 10 percent, water consumption was increasing by an average of 17 percent every year.[1] What frustrated Chesbrough was his firm belief that this dramatic increase in water consumption, an increase that became exponential after the opening of the lake tunnel, was being greatly exacerbated by waste. Every year in his reports—from 1871 until his departure, in 1879, from what had by then become the Department of Public Works—he stressed the need to find ways of conserving water. It was a noble effort, but it was not to succeed in his lifetime.

The second lake tunnel to the Two-Mile Crib went into service in 1875. Thanks to the new tunnel—along with the four engines at the Chicago Avenue pumping station and the distribution system's extension to the new pumping station on the west side, which came on line in 1877 with its additional four engines—the city's capacity for providing clean water was keeping pace with growth, despite the continuing problems with waste. Also during this period, the old wooden structure for the Two-Mile Crib was replaced with a masonry structure, and a proper lighthouse was added.

Although the deepening of the Illinois and Michigan Canal had managed to cleanse, relatively speaking, the south branch of the Chicago River, the condition of the river's north branch continued to worsen. For many years, Chesbrough and the Board of Public Works had discussed but never acted on a proposal to build a flushing canal or conduit from the lake to the north branch of the river, with pumps that would allow either for pumping lake water into the north branch or for pumping water from the north branch into the lake. Chesbrough was able to oversee the building of this structure, the Fullerton Avenue Conduit, through 1879. The conduit went into service the following year.

In January 1879, Chesbrough, over his protestations, was named commissioner of the Department of Public Works. This was not a happy time for his relationship with the city. For many years he had earned an annual salary of $3,500 as city engineer, but in 1877, in a meaningless gesture aimed at addressing the city's financial problems, his annual pay was reduced to $3,350. The final insult came after he was appointed commissioner and, in March 1879, saw his pay further reduced to $3,000. He'd had enough, and so, after twenty-four years with the city, he announced that his last day as commissioner would be May 18, 1879. The *Chicago Tribune*, echoing popular sentiment, called Chesbrough's imminent resignation "an embarrassment to the city" and a "personal as well as a public loss."[2]

Chesbrough, already in considerable demand as a consultant, was hardly at a loss for work. While he was still city engineer, and also during his later employment with the City of Chicago, numerous cities had called on him for his expertise. Now, working solely as a consultant, he was responsible for helping to plan sewerage systems for New Haven, Milwaukee, and Indianapolis. He also consulted on sewerage for the cities of Providence, Memphis, Peoria, Dubuque, Des Moines, Burlington (Iowa), Chattanooga, Pullman, Denver, Buffalo, and Winnipeg, and he served on the Sewerage Commission for Boston. He planned waterworks for Pittsburgh, Milwaukee, Jacksonville, and Akron, and he consulted on the water supplies of Boston, Cambridge, Toronto, Detroit, Albany, and Memphis. He also consulted for the Canada Southern Railroad Company on a proposed railroad tunnel under the Detroit River that would connect Canada with the United States. This tunnel, though not built during Chesbrough's lifetime, was a direct result of his success with the Chicago lake tunnel.

In 1880, Chesbrough was appointed permanent consulting engineer by the City of New York for the new Croton water supply, and, in association with Isaac Newton, chief engineer of the aqueduct, he planned the Quaker Bridge Dam and a new aqueduct. In 1882, the City of New York sent him to Europe to examine some high masonry dams in France and Spain, in connection with his work on the Quaker Bridge Dam. During that trip, he fell ill. He recovered sufficiently to return to New York in May 1883, and he was able to report the results of his examinations to New York's Department of Public Works, but he was never again a well man. He returned home to Chicago, and from that time on he was unable to work. He died on August 18, 1886.

A man of seemingly boundless energy, Chesbrough, in addition to his public service, had been one of the organizers in 1869 of the Civil Engineers'

Club of the Northwest, which in 1880 became the Western Society of Engineers, and he served as its president. He became a member of the American Society of Civil Engineers in June 1868 and was one of its directors in 1869. He served as the society's president from 1877 to 1878.

Chesbrough's legacy, for Chicago and for the nation as a whole, is hard to overstate. Chesbrough was largely responsible for bringing modern, systematic thinking to the problems of sewerage and water supply. It is difficult to imagine what might have become of Chicago if he had decided to stay in Boston in 1855. Perhaps another man would have come along with the vision, skill, and perseverance to accomplish what Chesbrough did. What is known, however, is that it was Ellis Sylvester Chesbrough who saved Chicago from the hazards of its own success. It was his sewerage system that turned the tide on the scourge of cholera; his boldness that secured for the city its first and enduring supply of safe water; his foresight that laid the foundation for the Sanitary and Ship Canal, widely hailed as one of the greatest engineering feats of the twentieth century. Experience gained on the lake tunnel also extended beyond Chicago. James Dull and James Gowen, because of their work on the lake tunnel, were chosen to assist in the building of the Hoosac Tunnel.[3] Similarly, the city of Cleveland relied on the Chesbrough system in building a water supply tunnel under Lake Erie in 1869–1874.[4]

* * *

On June 8, 1903, a discovery was made in the 50-foot westward extension from the land shaft that had been constructed to allow for placement of a transit to aid in digging the lake tunnel. That extension had been walled off before the tunnel's 1867 opening, but now, some thirty-five years later, workmen discovered a copper cylinder 18 inches long and 4 inches in diameter. They took the cylinder to Chief Engineer William Burns, who in turn notified City Engineer John Ericson and Assistant Engineer John Spengler. The cylinder was opened, and its contents were dried out and taken to Mayor Carter Harrison. Among the various items inside were paper currency, coins, and stamps from 1867; various newspaper accounts from January to March 1867 concerning the lake tunnel; a copy of the water commissioners' semiannual report from 1861; and reports from the Board of Public Works for 1864, 1865, and 1866. The cylinder also contained photographs of various people who had been involved in the lake tunnel project, among them William Clarke. The photograph of Clarke was of special interest because it included the following handwritten inscription on the back: "Length of tunnel by triangle as located equals 10,567.58 feet; length of tunnel by measure equals 10,567 feet; error equals .58 foot; center lines of tunnel at 8,270 feet from shore shaft, the point of meeting of the work from both ends, were found to be eight inches apart, the line from the east being the most southerly."[5] It seems that Clarke and the others were duly proud of their accomplishments and wanted to leave a memento for future generations.

In 1933, as part of the city's celebration of a century of progress, a ceremony was held to place a commemorative tablet at the Chicago Avenue water tower in honor of Ellis Chesbrough's contributions. Also honored were DeWitt Cregier and John Ericson, who had served the city in various capacities from 1884 to 1927. More than two hundred engineers attended the ceremony.

The Chicago lake tunnel and the Two-Mile Crib became templates for the city's growing water supply system, and their influence continues to the present day. From 1889 to 1935, seven more cribs were built in the lake, with associated tunnels providing the city with water (see table 4). The longest tunnel was the one to the Four-Mile Crib. Today two of the original eight cribs, the William Dever Crib and the Edward Dunne Crib, remain in operation, providing the city and surrounding communities with fresh water.[6]

Table 4. History and status of lake tunnel's cribs

NAME	CONSTRUCTION STARTED	PLACED IN SERVICE	STATUS IN 2016	APPROXIMATE DISTANCE FROM SHORE (IN MI., MEASURED 1937)	WATER DEPTH (IN MI.)
Two-Mile Crib	1864	1867	Demolished (1936)	2	29
Four-Mile Crib	1889	1892	Deactivated	3.6	37
68th Street Crib	1892	1894	Deactivated	2	32
Lake View Crib	1889	1895	Demolished (1923)	1.9	——*
Carter Harrison Crib	1897	1900	Tunnel collapsed (1997)	2.6	33.8
Edward Dunne Crib	1907	1911	In service	2	32
Wilson Avenue Crib	1915	1918	Deactivated	2.1	33.5
William Dever Crib	1927	1935	In service	2.6	34

* Depth of Lake View Crib not reported in source material.

Source: Civil News Service of Illinois, *A Pictorial Story of Chicago's Public Works: Engineering Wonders of the World, Featuring Chicago's Charter Jubilee, 1837–1937* (Chicago: Civil News Service of Illinois, 1937), 45.

Figure 41. Demolition of the Two-Mile Crib, 1936. Source: Historical Collections of the Great Lakes, Jerome Library, Bowling Green State University.

On February 6, 1936, the gate to the great Chicago lake tunnel at the Two-Mile Crib was closed so that the tunnel could be dewatered. In April of that year, the contractor Charles Fitzsimmons and the C. J. Connell Dredging Company began filling the lake shaft with concrete, after the tunnel gate had been reopened so that the concrete could flow into the tunnel. On August 28, once the shaft was sealed, demolition of the crib itself began.[7]

On September 21, at three o'clock in the afternoon, WENR, a local radio station broadcasting from the US Coast Guard cutter *Rush*, transmitted the story and sound of the final blasting of the crib. The contractor, by agreement with the city, had begun dredging on September 15 and would continue until October 21, working on the area surrounding the crib to a depth of 30 feet below the surface of the lake, and removing all the remaining materials from the crib.[8] And so, after almost seventy years from the time its excavation was completed, the great Chicago lake tunnel was no more. Its legacy, however, and the long shadow of the extraordinary work of Ellis Sylvester Chesbrough, will live on so long as the cool waters of Lake Michigan remain to quench a city's thirst.

ACKNOWLEDGMENTS

I am indebted to a number of people for their assistance in the conception and implementation of this book. First, to my many sailing students and friends who over the years have found the existing water cribs to be objects of mystery, interest, and beauty—it was your curiosity and questions that led me to this work.

I have also been greatly assisted by the kind and competent work of the following research librarians, archivists, and curators: Marc Arenberg and Crispien Van Aelst at the Illinois Regional Archive Depository; Glenn E. Humphreys at Harold Washington Library; Ginger Frere at the Newberry Library; Leslie Martin, Joy Bivins, Ellen Keith, Trenton Carls, Michael Featherstone, and Sarah Yarrito at the Chicago Historical Museum; Caitlin Lampman at the Arizona Historical Society; Bob Graham at the Jerome Library, Bowling Green State University; Dawn Roberts at the Chicago Academy of Sciences; Wayne Lorentz of the Chicago Architecture Blog; Michael Frost at the Sterling Memorial Library, Yale University; David R. Phillips of the David R. Phillips Collection; Paul Grochowski of the University of Michigan Art, Architecture, and Engineering Library; Albert Quattrucci, Rachel Waters, and Susan Stravinski of the University of Wisconsin–Madison Memorial Library; Pamela Priest and Sarah Kozma of the Onondaga Historical Association; and Diana Rose-Harper, Christine Colburn, and Eileen Ielmini of the Special Collections Research Center at the University of Chicago. Thank you to Pamela Bannos for valuable information about the changing lakefront of Chicago, to Christopher Burke and Cliff Schexnayder for helping me better understand the engineering of the lake tunnel, to Laura Burke and Orion Galey for assistance with the archives of the Western Society of Engineers, and to Ed Lyons for his advice and encouragement. I am also indebted to David Solzman for his review and comments on the text, and to Gianna Francesca Mosser and Xavier Callahan of Northwestern University Press for their editorial advice and guidance.

And for her counsel, careful reading, editorial advice, and abiding patience in listening to my endless talk about the great Chicago lake tunnel, I thank my wife, Jill Mateo—the wonder of my world.

APPENDIXES

Appendix A: Population of Early Chicago, 1833–69

YEAR	POPULATION	YEAR	POPULATION
1833	350	1852	38,734
1834	2,000	1853	60,662
1835	3,265	1854	65,872
1836	3,820	1855	80,023
1837	4,180	1856	84,113
1838	4,000	1857	93,000
1839	4,200	1858	90,000
1840	4,479	1859	95,000
1841	5,500	1860	109,206
1842	6,590	1861	120,000
1843	7,580	1862	138,186
1844	8,000	1863	160,000
1845	12,088	1864	160,353
1846	14,169	1865	178,492
1847	16,859	1866	200,418
1848	20,023	1867	220,000
1849	23,047	1868	252,054
1850	28,269	1869	273,043
1851	34,000		

Sources: Homer Hoyt, *One Hundred Years of Land Values in Chicago: The Relationship of the Growth of Chicago to the Rise of Its Land Values, 1830–1933* (Chicago: University of Chicago Press, 1933), 19 (1833–36); Alfred T. Andreas, *History of Chicago, from the Earliest Period to the Present Time*, 3 vols. (Chicago: Alfred T. Andreas, 1884–1886), 1:183 (1837–57), 2:52 (1858–69).

Appendix B: Chicago Political Roster, 1837–71

The following information is taken from Alfred T. Andreas, *History of Chicago, from the Earliest Period to the Present Time*, 3 vols. (Chicago: Alfred T. Andreas, 1884–1886), 2:49–50 (mayors, 1837–61; elected officials, 1861–71) and from City of Chicago, *First Annual Report of the Department of Public Works to the City Council of the City of Chicago for the Fiscal Year Ending December 31, 1876* (Chicago: Clark & Edwards, 1877), 2 (water commissioners, 1851–61; sewerage commissioners, 1855–61; public works commissioners, 1861–71).

Mayors, 1837–61

William Ogden (1837–38)
Buckner Morris (1838–39)
Benjamin Raymond (1839–40)
Alexander Loyd (1840–41)
Francis Sherman (1841–42)
Benjamin Raymond (1842–43)
Augustus Garrett (1843–44)
Alson Sherman (1844)
Augustus Garrett (1845–46)
John Chapin (1846–47)
James Curtiss (1847)
James Woodworth (1848–49)
James Curtiss (1850)
Walter Gurnee (1851–52)
Charles Gray (1853)
Isaac Milliken (1854–55)
Levi Boone (1855)
Thomas Dyer (1856–57)
John Wentworth (1857–58)
John Haines (1858–60)
John Wentworth (1860–61)

Elected Officials, 1861–71

1861

Mayor: Julian S. Rumsey
City clerk: A. J. Marble
City attorney: Ira W. Buel
City treasurer: W. H. Rice
Aldermen by wards: (1) J. K. Botsford, William Colby; (2) J. Q. Hoyt, James M. Marshall; (3) A. D. Titsworth, Hiram Joy; (4) William Baragwanath, Samuel Myers; (5) C. C. P. Holden, Robert H. Foss; (6) Edward S. Salomon, James W. Cobb; (7) Alonzo Harvey,* Gurdon S. Hubbard; (8) W. G. White, Redmond Prindiville; (9) Robert Law, Gurdon Perkins; (10) John Comiskey, Malcolm McDonald

*Andrew Harvey was first declared elected, but the council afterward reconsidered its action, and Alonzo Harvey was declared chosen on June 17.

1862

Mayor: F. C. Sherman
City clerk: A. Marble
City attorney: George A. Meech
City treasurer: William H. Rice*
Aldermen by wards: (1) John T. Edwards, J. K. Botsford; (2) Peter Shimp, J. Q. Hoyt; (3) James A. Hahn, A. D. Titsworth; (4) Andrew Schall, William Baragwanath; (5) William A. Groves, C. C. P. Holden; (6) Francis C. Brown, Edward S. Salomon; (7) James Conlan, Alonzo Harvey; (8) Charles L. Woodman, W. G. White; (9) William T. Shufeldt, Robert Law; (10) Redmond Sheridan, John Comiskey

* F. H. Cutting was elected but failed to qualify, and a new bond was filed by William H. Rice, the incumbent.

1863

Mayor: F. C. Sherman
City clerk: Henry W. Zimmerman
City attorney: Francis Adams
City treasurer: David A. Gage
Aldermen by wards: (1) James A. Hahn, Andrew Schall; (2) A. D. Titsworth, Peter Shimp; (3) James H. Roberts, Stephen Barrett; (4) Benjamin E. Gallup, John T. Edwards; (5) Constantine Kahn, Mark Sheridan; (6) David Walsh, Malcolm McDonald; (7) James E. Abbott, John Comiskey; (8) Richard Clark, Redmond Sheridan;* (9) Mancel Talcott, Francis C. Brown; (10) George Himrod, C. C. P. Holden; (11) George Von Hollen, L. L. Bond; (12) William Gastfield, Christian Casselman; (13) John M. Armstrong, David Aleckner; (14) Valentine Ruh, Anton Hottinger; (15) Michael Sullivan, James Conlan; (16) William T. Shufeldt, C. L. Woodman

*Francis J. Ullbrich elected to fill the vacancy caused by Redmond Sheridan's resignation.

1864

Mayor: F. C. Sherman
City clerk: Henry W. Zimmerman
City attorney: Francis Adams
City treasurer: David A. Gage
Aldermen by wards: (1) George W. Gage, Charles D. Peacock; (2) Peter Shimp, A. D. Titsworth; (3) Stephen Barrett, James H. Roberts; (4) Samuel McRoy,

Benjamin E. Gallup; (5) Mark Sheridan, Constantine Kahn; (6) John Wallwork, David Walsh; (7) Joseph Sherwin, John Comiskey; (8) Patrick Rafferty, Richard Clark; (9) Willard Woodard, Mancel Talcott; (10) C. C. P. Holden, George Himrod; (11) Lester L. Bond, George Von Hollen; (12) Nathaniel W. Huntley, William Gastfield; (13) Mathias Franzen, John M. Armstrong; (14) A. Hottinger, Valentine Ruh; (15) Iver Lawson, Michael Sullivan; (16) Charles L. Woodman, James J. O'Sullivan

1865

Mayor: John B. Rice
City clerk: Albert H. Bodman
City attorney: Daniel D. Driscoll
City treasurer: A. G. Throop
Aldermen by wards: (1) Joshua C. Knickerbocker, George W. Gage; (2) William H. Carter, Peter Shimp; (3) Charles G. Wicker, Stephen Barrett; (4) H. M. Willmarth, Samuel McRoy; (5) Constantine Kahn, Mark Sheridan; (6) Thomas C. Hatch, John Wallwork; (7) Avery Moore, Joseph Sherwin; (8) M. L. Frisbee, Patrick Rafferty; (9) Mancel Talcott, Willard Woodard; (10) Edward Bixby, C. C. P. Holden; (11) S. I. Russell, Lester L. Bond; (12) William Gastfield, Nathaniel W. Huntley; (13) L. Proudfoot, Mathias Franzen; (14) Valentine Ruh, A. Hottinger; (15) Samuel Shackford, Iver Lawson; (16) Robert Clark, Charles L. Woodman

1866

Mayor: John B. Rice
City clerk: Albert H. Bodman
City attorney: Daniel D. Driscoll
City treasurer: A. G. Throop
Aldermen by wards: (1) William Cox, J. C. Knickerbocker; (2) Calvin DeWolf, William H. Carter; (3) Stephen Barrett, Charles G. Wicker; (4) Allen C. Calkins, H. M. Willmarth; (5) M. Finucan, Constantine Kahn; (6) John Wallwork, Thomas C. Hatch; (7) Max Schuler, Avery Moore; (8) Patrick Rafferty, M. L. Frisbee; (9) Willard Woodard, Mancel Talcott; (10) C. C. P. Holden, Edward Bixby; (11) Henry Ackhoff, S. I Russell; (12) N. W. Huntley, William Gastfield; (13) M. Franzen, L. Proudfoot; (14) Robert Engel, Valentine Ruh; (15) Iver Lawson; Samuel Shackford; (16) J. J. O'Sullivan,* Robert Clark

*Michael O'Sullivan elected to fill the vacancy caused by J. J. O'Sullivan's resignation.

1867

Mayor: John B. Rice
City clerk: A. H. Bodman
City attorney: Hasbrouck Davis
City treasurer: William F. Wentworth
Aldermen by wards: (1) Joshua C. Knickerbocker, William Cox; (2) Arthur Dixon, Calvin DeWolf; (3) Charles G. Wicker, Stephen Barrett; (4) Samuel McRoy, Allen C. Calkins; (5) John Raber, M. Finucan; (6) David Walsh, John Wallwork; (7) John MacAllister, Max Schuler; (8) John Comiskey, Patrick Rafferty; (9) John H. Carpenter, Willard Woodard; (10) Edmund Bixby,* C. C. P. Holden; (11) S. I. Russell, Henry Ackhoff; (12) C. J. Casselman, N. W. Huntley;** (13) George T. Beebe, M. Franzen; (14) Theodore Schintz, Robert Engel;† (15) Samuel Shackford, Iver Lawson; (16) George B. Mansur, M. O'Sullivan

*Died December 5.
**John Buehler elected to fill the vacancy caused by N. W. Huntley's resignation.
†Resigned December 2.

1868

Mayor: John B. Rice
City clerk: A. H. Bodman
City attorney: Hasbrouck Davis
City treasurer: W. F. Wentworth
Aldermen by wards: (1) William Cox, Joshua C. Knickerbocker; (2) P. M. Donnellan, Arthur Dixon; (3) Stephen Barrett,* Charles G. Wicker; (4) A. C. Calkins, Samuel McRoy; (5) Mark Sheridan, John Raber; (6) Michael Keeley, David Walsh; (7) James H. Hildreth, John MacAllister; (8) Patrick Rafferty, John Comiskey; (9) Willard Woodard, John H. Carpenter; (10) C. C. P. Holden, Alvin Salisbury; (11) B. F. Russell, S. I. Russell; (12) John Buehler, C. J. Casselman; (13) K. G. Schmidt, George T. Beebe; (14) Louis A. Berger, Theodore Schintz; (15) John Herting, Samuel Shackford; (16) Edward Kehoe, George B. Mansur

*Died May 21, and James A. Hahn elected to fill the vacancy.

1869

Mayor: John B. Rice
City clerk: A. H. Bodman
City attorney: Hasbrouck Davis
City treasurer: W. F. Wentworth
Aldermen by wards: (1) William Cox, Joshua C. Knickerbocker; (2) Patrick M. Donnellan, Arthur Dixon; (3) James A. Hahn, Charles G. Wicker; (4) A. C. Calkins, Samuel McRoy; (5) Mark Sheridan, John Raber; (6) Michael Keeley, David Walsh; (7) James H. Hildreth, John MacAllister; (8) Patrick Rafferty, John Comiskey; (9) Willard Woodard, John H. Carpenter; (10) C. C. P. Holden, Alvin Salisbury; (11) B. F. Russell, S. I. Russell; (12) John Buehler, C. J. Casselman; (13) K. G. Schmidt, George T. Beebe; (14) Louis A. Berger, Theodore Schintz; (15) John Herting, Samuel Shackford; (16) Edward Kehoe, George B. Mansur

*On March 10, the city was divided into twenty wards, and the time for the city election changed from April to November; incumbents were continued until the first Monday of December.

1869–70

Mayor: R. B. Mason*
City clerk: Charles T. Hetchkiss
City attorney: Israel N. Stiles
City treasurer: David A. Gage
Aldermen by wards: (1) Richard Somers, William Cox; (2) Arthur Dixon, P. M. Donnellan; (3) Joseph A. Montgomery, James A. Hahn; (4) John H. McAvoy, A. C. Calkins; (5) George S. Whitaker, Peter Daggy; (6) William Tracey, Mark Sheridan;** (7) William Batterman, P. J. Hickey; (8) William S. Powell, James H. Hildreth; (9) George Powell, John Comiskey; (10) Thomas Wilce, C. C. P. Holden; (11) James Walsh, B. F. Russell; (12) Samuel McCotter, Willard Woodard; (13) James L. Campbell, A. D. Robinson; (14) P. B. Shiel, B. G. Gill; (15) James J. McGrath, John Buehler; (16) James D. Tyler, K. G. Schmidt; (17) Theodore Schintz, Louis A. Berger; (18) Thomas Carney, A. Bengley; (19) James McCauley, John Herting; (20) M. A. Devine, Edward Kehoe

*Elected in November (on March 10, 1869, the time for the city election changed from April to November, with incumbents continued until the first Monday of December; on the same date, the city was divided into twenty wards).
**Daniel Heenan elected to fill the vacancy caused by Mark Sheridan's resignation.

1870–71

Mayor: R. B. Mason
City clerk: Charles T. Hotchkiss
City attorney: I. N. Stiles
City treasurer: David A. Gage
Aldermen by wards: (1) John J. Knickerbocker, Richard Somers; (2) Joseph E. Otis, Arthur Dixon; (3) D. Coey, Joseph A. Montgomery; (4) Harvey M. Thompson, John H. McAvoy; (5) Peter Daggy, George S. Whitaker; (6) Michael Schmitz, William Tracey; (7) P. J. Mickey, William Batterman; (8) Michael B. Bailey, William S. Powell; (9) William B. Bateham, George Powell; (10) C. C. P. Holden, Thomas Wilce; (11) Herman O. Glade, James Walsh; (12) Henry Witbeck, Samuel McCotter; (13) S. S. Gardner, James L. Campbell; (14) B. G. Gill, P. B. Shiel; (15) John Buehler, James J. McGrath; (16) K. G. Schmidt, James D. Tyler; (17) Louis Schaffner, Theodore Schintz; (18) John McCaffrey, Thomas Carney; (19) William M. Clarke, James McCauley; (20) Gustavus A. Busse, M. A. Devine

Water Commissioners, 1851–61

1851–54: Horatio G. Loomis,* Alson S. Sherman, John B. Turner [P]
1854–56: J. H. Woodworth [P], John C. Haines, George W. Dole
1856–58: Orrington Lunt, John C. Haines, George W. Dole [P]
1858–60: Orrington Lunt, Noah Sturtevant, George W. Dole [P]
1860–61: Orrington Lunt [P], Edward Hamilton, Benjamin Carpenter

*Names ordered by south, west, and north divisions, respectively.
**[P] denotes *president*.

Sewerage Commissioners, 1855–61

1855–59: Joseph D. Webster,* Sylvester Lind, William B. Ogden [P]
1859–61: Joseph D. Webster [P], Sylvester Lind, Philip Conley

*Names ordered by south, west, and north divisions, respectively.
**[P] denotes *president*.

Public Works Commissioners, 1861–71

1861*–63: John G. Gindele,** Frederick Letz, Benjamin Carpenter [P]†
1863–65: Mayor Francis C. Sherman, ex officio member
1863–67: John G. Gindele [P], Frederick Letz, Orrin J. Rose
1867–69: William H. Carter, John McArthur, Augustus H. Burley [P]
1869–71: William H. Carter, John McArthur [P], Redmond Prindiville
1865–69: William Gooding, Roswell B. Mason‡
1869–71: William Gooding, Edward B. Talcott‡

*The Board of Water Commissioners and the Board of Sewerage Commissioners were merged on May 6, 1861, into the Board of Public Works.
**Names ordered by south, west, and north divisions, respectively.
†[P] denotes *president*.
‡Members of the board in matters pertaining to the cleansing of the Chicago River.

Appendix C: Annotated Contract and Specifications for the Lake Tunnel

In the following material, square brackets indicate annotations to the contract as reported in William Blake, *Civil Engineering and Public Works, Paris Universal Exposition 1867: Reports of the United States Commissioners* (Washington, DC: Government Printing Office, 1870), 19–25.

LOCATION AND GENERAL DESCRIPTION. The tunnel is to commence at such point as may be selected by the board of public works, on the lot now occupied by the pumping works of the city of Chicago, at the east end of Chicago avenue, and on the shore of Lake Michigan; and to extend two miles out under the lake, in a straight line, at right angles to the general direction of the shore.

The bottom of the inside surface of the east end of the tunnel shall be sixty-six feet below the ordinary level of the lake, or sixty-four feet below what is usually known as "City Datum"; and the bottom surface shall descend uniformly at the rate of two feet per mile to the west end of the tunnel.

There are to be one land and two to four lake shafts; the land shaft at the west end, one lake shaft at the east end, and the remaining lake shaft or shafts at such intermediate points as shall be determined upon by said board, when the proper time for locating them shall arrive. The lake shafts are to consist of cast iron cylinders, and to be protected by hollow pentagonal cribs. [Only one lake shaft and one crib were constructed.]

The tunnel is to be very nearly circular in form, and to have an interior width of five feet, and height of five feet and two inches.

CRIBS. It has been proposed to construct four cribs, on the supposition that this number might be required to complete the tunnel in two years, but if, after commencing the work, it shall be found in time that one or more of them may be omitted, such omission shall be made; and for this reason but two cribs [Only one crib, the outer one, was found necessary.] shall be commenced before the probable rate of progress in the tunnel, from the land shaft, shall have been satisfactorily ascertained.

DESCRIPTION. The cribs are to be five-sided, each outer side to be fifty-eight feet long. There is to be a central space in each crib, in form similar to the outside, leaving the thickness between the central space and outside of the crib twenty-five feet. The interior and exterior sides of the cribs are to be perpendicular from their bottoms to their tops, which are to be five feet above the ordinary surface of the lake; hence the outermost crib is to be forty feet high. The height of the others will depend upon their location, which will be determined during the progress of the work.

The outer crib is to have three openings through its sides, one opening through each of the western and southern sides; each opening to be five feet high and four feet wide, and to be connected with the top of the crib by a well four feet square. The most northern opening is to have its bottom five feet above the bottom of the crib, the middle one eleven feet higher, and the southern one eleven feet higher than the middle. Each opening is to be furnished with an iron paddle gate, to be worked by means of an iron rod, from the top of the crib, as shown on plans.

Each crib is to be provided with two gates, [This was afterward changed, and eight smaller and circular gates were constructed so as to regulate the flow of the water into eight watertight compartments, which occupy the interior of the crib.] each two feet high, and one and a half foot wide, placed five feet above the bottom of the crib, and against the wall of the inside space. These gates are to be used for regulating the sinking of the cribs to their places, and worked by rods from the top of the crib, as shown on plans. Wrought-iron gratings to protect these gates must be provided, as shown on these plans.

MODE OF CONSTRUCTION. The bottom of each crib shall be formed by an outer, inner, and middle line of twelve-inch-square white pine timber, which shall be connected with cross-timbers of the same size, and with twelve-by-three-inch joists placed two feet apart from center to center, and the underside of the whole, except the central space of the crib, sheathed over with two-inch pine plank, fastened on by six-inch spikes, driven through each plank in every timber and joist, as shown on plan. [Changed in construction to twelve-inch timbers.] All the timbers and cross-timbers are to be connected by dovetail joints the full width of each timber, and of equal depth in each, the dovetail being one inch deeper at its end than at its shoulder. On the outer, inner, and middle lines of timber, and on the angle pieces of the outer and middle lines, solid walls of twelve-inch-square timber are to be built up to a height of forty feet above the bottom for the outer crib, and to a height of five feet above ordinary watermark for the other cribs. The middle wall extends solid from outside to outside of the crib; the inner wall only around the inner space. The angle timbers are eleven and a half feet long from tip to tip. Between the outer and middle walls there are two lines of cross-timbers on each side of the crib, and between the middle and inner walls three lines. [In construction, these timbers were placed so that each piece extended from the outer through the middle to the inner wall.] The cross-timbers are twelve inches square, placed twelve inches apart above each other, and extend through each wall. All of the timber and lumber, except the upper twelve feet in each outside wall, is to be of sound white pine. The upper twelve feet of the outer wall to be of sound white oak, free from sap, or any imperfections tending to hasten decay. All of the joints of the timbers at the angles of the outer and inner walls, at the ends of the middle walls, and at the ends of the cross-timbers are to be dovetailed like those described for the bottom course of timbers in the crib [at the corners]. Where the timbers of the middle wall cross each other, they are to be notched half and half. Whenever the ends of timbers butt against each other in the outer and inner walls, they must do so at the center of the end of some cross-timber, where the dovetailed end can be made to lap over the butt joint. On the bottoms and sides of the three openings through the outer crib, twelve-inch-square timbers touching each other are to be run entirely through the ends; and across the top of the openings six-inch planks are to be spiked, fitting close to each other. The walls from each opening to the top of the crib are to be formed of six-inch planks placed horizontally, and notched into the crib timbers wherever they can be.

The timbers in the outer and middle walls of each crib are to be fastened to each other by 1 5/16-inch-square wrought-iron bolts thirty inches long, with ragged edges, and driven into the timber at an angle twelve degrees from a perpendicular, and inclining alternately toward each other. [The ragging was omitted.] The timbers of the inner wall are to be fastened in a similar manner to those of the outer and middle, except that the bolts are to be but one inch square.

Each outer angle of the crib is to be protected by a covering of wrought iron one inch thick, extending two feet each way from the corner, and from the top of the crib downward ten feet. These angle irons are to be fastened on by round wrought-iron one-and-a-half-inch bolts, one through every timber, on each side of the corner. The bolts are to be alternately long and short, the long being fastened through the middle timbers of the crib, and the short through the outer timbers, as shown on plans. The long bolts are to be used also in the lower part of the crib, from the bottom to the angle iron, in every other timber. [On each side of each angle, from the bottom to the top of the crib, and in place of the sheeting, a piece of oak 12 by 14 inches was put in the line of the long bolts. It was cut down to 3 inches under the angle irons. A corresponding piece of oak, 3 inches wide all the way, was put at the other ends of the bolts on the middle wall.]

The spaces between the joists at the bottom of the crib are to be filled with gravel or broken stone flush with the tops of the joists. The joists are to be planked over with two-inch pine, fastened down with six- inch spikes. [This was all made of 12-inch timber.]

There are to be twenty-five pair of one-inch round wrought-iron rods to connect saddles placed under bottom timbers and over top ones, as shown on plans. The forms and mode of placing the saddles and connecting the rods with them and with each other are shown on plans. [These were not put in.]

The whole of the joints on the bottom of the crib, around the outside and central space of the crib from its bottom to within four feet of the top, and the joints around the inside of the three openings and their wells, are to be thoroughly caulked with oakum and paid with tar. [Every space from the bottom to the top of the crib, including the middle wall, was caulked.]

Each crib shall, immediately after being towed to its proper position, be secured by one-and-a-half-inch iron cables to five Mitchell's mooring screws, forced ten feet into the clay at the bottom of the lake. The dimensions and form of these screws can be seen on plans.

Each crib is to be filled with sound rubble stone, from its bottom to its top, as soon after being moored as practicable.

The angles of each crib shall be placed exactly in such positions as the board of public works may direct.

THE LAND SHAFT. This will be located near and in the rear of the present pumping works of this city. From the surface of the ground to a depth of fifteen feet below the level of the lake, the shaft is to be twelve feet in diameter; and then it is to be contracted by a sloping offset of three feet all around to six feet diameter, to five feet below the bottom of the invert of the tunnel, a distance of sixty feet. [The shaft was not executed in this way. Iron cylinders, three in number and extending to a depth of 30 feet, were substituted. These were 9 feet in diameter and an inch and a half thick. The shaft below was made 8 feet clear diameter and 12 inches thick.]

The whole of the shaft is to be lined with brick masonry twelve inches thick, where the shaft is twelve feet in diameter, and eight inches thick, where it is six feet in diameter, and on the invert at its bottom. The masonry of the offset is to be twelve inches thick.

The masonry is to be formed of hard-burned, clear-ringing, and well-formed bricks, entirely free from lime, not less than eight inches long, two and a quarter inches thick, and four inches wide, to be laid upright in cement mortar, and in rings or shells four inches thick. The courses must be horizontal, the inside surface of the shaft must be true and cylindrical, and the joints between the bricks not over a quarter of an inch on the inside of each ring.

The cement mortar is to be equal to the best of Clark's LaSalle, and mixed with one measure of clean sharp sand to one of cement, and used as soon as possible after being mixed.

The joints between the rings must be not less than half an inch, and all the joints in the masonry must be perfectly filled with mortar at the time the masonry is laid.

Where the natural soil around the shaft is sand or loose material, there is to be not less than twelve inches of thoroughly puddled clay on the outside of the masonry. [All spaces were filled with either concrete or brickwork in cement.]

Whatever timber and lumber may be necessary to support the earth around the shaft, before the masonry can be built, must be furnished and put in by the contractor.

The pumping, and all labor and machinery connected therewith, must be done and furnished by the contractor.

INLET CYLINDER FOR OUTER LAKE SHAFT. To be nine feet, in diameter inside, and two and a quarter inches thick. To be made in sections of about nine feet in length, so that seven sections will make up the total length of the cylinder, or sixty-four feet. The flanges to be five and five-eighths inches wide by two and a quarter inches thick, and to be faced in the lathe true and at right angles to the center line of the cylinder. Each flange to have a small annular groove turned into the face, to receive the putty to be used in making the joint, as shown on the drawing. Flanges to be drilled for bolts one and a half inch in diameter and about seven inches between centers. The lower section to be turned a taper of one-sixteenth of an inch, in nine inches from the end on the outside. The balance of the section to be accurately and truly turned parallel the entire length above the taper.

The second and third sections, and, if required, the fourth section, from the bottom to be accurately and smoothly turned, [Only the lowest section was turned, the castings proving sufficiently smooth for the others without turning.] each section of a diameter slightly in excess of the one next below it, say the sixty-fourth part of an inch, and should there be any difference in the diameter of opposite ends of the same section, then the largest end to be connected uppermost.

The section next below the top one is to be provided with openings for inlet gates, plans and specifications for which will be given to the contractor whenever necessary after one month from the letting of the work, and all additional expense to the contractor in making such opening will be paid for as extra.

INTERMEDIATE CYLINDERS FOR LAKE SHAFTS. To be one to three in number, to be nine feet in diameter inside, and one and five-eighths inch thick. To be made in sections of about nine feet in length, so that seven sections will make up the total length of cylinder, whether sixty-five, sixty-six, or sixty-seven feet. The flanges to be five inches wide by one and three-quarter inch thick, and faced as specified for the "inlet cylinder." Flanges to be drilled for bolts one and a half inch in diameter, and about seven inches between centers. Three or four of the lower sections to be turned on the outside as may be required, and as specified for the "inlet cylinder." The sections to be stiffened by two intermediate internal flanges two and three-eighths inches wide by one and three-quarter inch thick. At the top of the third or fourth section from the bottom, as may be required, the flange to have a double width to receive a bonnet or head for closing and disconnecting the cylinder at that joint, as delineated on the drawing.

The iron employed in the construction of the cylinders to be of good quality. The castings to be sound and free from sand holes or other defects.

Bolts for connecting the cylinders to be made of the best wrought iron, one and a half inch in diameter, with hexagonal heads and nuts.

Joints between cylinder section flanges to be made airtight with thin red lead putty. All necessary work, materials, machinery, and tools for completing and putting in place the cribs and cylinders to be done and furnished by the contractor. [The intermediate cylinders and shafts thus provided for were not found necessary and were not put in.]

THE TUNNEL PROPER. The clear width of the tunnel is to be five feet, and the clear height five feet and two inches, the top and bottom arches to be semicircles. The tunnel is to be lined with brick masonry eight inches thick, in two rings or shells, the bricks to be laid lengthwise of the tunnel with toothing joints. The mortar, character of materials, and workmanship are to be like those described for the land shaft.

The excavation for the tunnel, when through sufficiently firm clay, shall conform exactly to the outside of the masonry on the bottom and sides. On top, just enough of excavation above the masonry will be allowed, to give room to turn and key the upper arch properly. [In executing this part of the work, the masons soon learned to put in the masonry in clay ground without removing any more earth than was necessary for the regular arch.] Sections of not over two feet of the upper arch are to be built at a time, and immediately after each section is keyed, the space above it shall be filled with earth, which shall be put into the space in small quantities at a time, and thoroughly rammed until it becomes as solid as the natural soil above.

Should a soil be met with in any part of the tunnel requiring the sides, bottom, and top to be planked and braced before the masonry can be built, filling in, carefully and thoroughly done, shall be put in between the masonry and the boarding, in or between the masonry and sides of the excavation, as is specified for the upper arch in a firm soil. The material for this filling may be dry sand or puddled clay. [All the lining around the masonry of the tunnel was made with brick and cement.]

All timber and lumber necessary for bracing and supporting the sides of the tunnel and shafts, previous to the completion of the masonry, must be furnished and put in place by the contractor.

The contractor must furnish and put in place all necessary airpipes and apparatus for ventilating the tunnel, all pumps, steam engines, hoisting apparatus and fixtures for the same, all sheds and shelters for the protection of workmen and materials on the cribs, and all necessary tracks, trucks, and other necessary implements for machinery for removing excavated material out of, and building materials into, the tunnel.

The contractor must also remove all excavated material taken out of each shaft to such points as shall be designated by the board of public works, provided said points shall not exceed three hundred feet from any shaft, and must provide all necessary tugboats, scows, and other means of transportation, and implements required for such removal.

GENERAL SPECIFICATIONS. The contractor shall furnish and maintain such lights as may be necessary to avoid danger to navigation; and in case of failure to do so, shall be liable for all damages the city may have to pay on account of such failure.

All materials, of whatever kind, to be used in the work, are to be inspected by the board of public works; and all unsuitable materials are to be immediately removed from the work by the contractor.

The contractor shall discharge from his employment, when directed by the board of public works, all unfaithful and incompetent workmen.

The board of public works must be permitted to remove such portions of the work as they may, from time to time, think necessary for the discovery of improper materials or workmanship; and the contractor shall restore such work at his own expense, in case it shall have been done improperly; and at the expense of said board if done in a proper manner.

Any work, materials, machinery, or tools necessary for the completion of the tunnel, cribs or shafts, omitted in the plan and specifications, shall be done or furnished by the contractor, and paid for as extra work, at such valuation as the board of public works may make. This section is not to apply to the completion of the work as specified, but only to extra work, and no tools and machinery to be paid for unless ordered by the board for extra work.

The contractor shall furnish men and stakes sufficient to enable the engineer in charge of the work to give the necessary lines and levels to construct the work by.

The contractor must deliver to the board of public works, on or before the first day of each month, a written statement of the amount of extra work done and extra materials furnished during the previous month.

The contractor will be required to keep the work in perfect repair for twelve months after the same shall have been faithfully completed to the satisfaction of the board of public works.

Monthly estimates will be made by the board of public works of the value of work actually done and in its permanent place; and on or about the sixth day of each month seventy-five per cent of the estimated value of the work done the previous month will be paid the contractor; the remaining twenty-five per cent being reserved as security for the faithful completion of the whole work. [During the progress of the work, this provision was changed by vote of the city council, and 85 per cent was paid upon monthly estimates for all work done, including the crib before it was launched.]

STATEMENT OF QUANTITIES OF MATERIAL IN THE OUTERMOST CRIB [the quantities of these materials were considerably increased in consequence of the changes made during the progress of the work]. 38,814 cubic feet of white pine timber; 3,500 cubic feet of white oak timber; 20,251 feet, board measure, of white pine two-inch plank; 1,700 wrought-iron 1 5/16-inch square bolts, 30 inches long; 120 wrought-iron 1-inch square bolts, 30 inches long;

200 wrought-iron 1 1/2-inch round bolts, 13 1/2 feet long; 50 wrought-iron 1 1/2-inch round bolts, 18 inches long; 200 square feet of 1-inch angle iron; 6,025 cubic yards of stone filling; besides 6-inch spikes, and paddle and other gates.

Appendix D: Expenditures on the Lake Tunnel and Crib, 1863–April 1867

1863	EXPENDITURES MINUS DEDUCTIONS ($)
Services of tugs in taking soundings	429.00
Labor	700.85
Draughtsman	339.00
Tools, blocks, etc.	72.96
Use of scows	408.00
Advertising for proposals, printing specifications, etc.	346.25
Legal services	125.00
Traveling expenses of engineer and others	195.84
Telegrams	56.46
Rope, iron, and blacksmith work	122.27
Watching scows, use of boat	42.00
Miscellaneous	31.15
Work at shop	50.84
TOTAL FOR 1863	2,919.62

1864	EXPENDITURES MINUS DEDUCTIONS ($)
Dull & Gowen, on account lake tunnel, outer crib, and extras	97,670.22
Services of engineer	250.46
Services of draughtsman	449.55
Services of inspectors	4,803.00
Labor	395.67
Lumber	804.21
Piles, including freight	718.49

1864	EXPENDITURES MINUS DEDUCTIONS ($)
Office fixtures (inspectors)	58.15
E. W. Smith, consulting engineer	400.00
Hardware	53.85
Brick	17.50
Repairing instruments	192.75
Chain	36.91
Horse keeping	53.65
Coal for inspectors' office	42.90
Miscellaneous	154.05
Labor at pumping works shop	287.88
TOTAL FOR 1864	106,389.24

1865	EXPENDITURES MINUS DEDUCTIONS ($)
Dull & Gowen, on account lake tunnel, outer crib, and extras	188,551.49
Services of engineer	1,075.00
Services of draughtsman	72.00
Services of inspectors	7,031.00
Lumber	54.27
Lumber for crib	338.51
Piles, including freight	338.00
Bedding and furniture for inspectors at crib	93.18
Castings for shaft at crib	11,923.09
Freight and hauling castings for shaft	706.77
Castings	265.39
Dredging for crib	1,500.00
Use of tugs	2,533.00

1865	EXPENDITURES MINUS DEDUCTIONS ($)
Removing piles	209.38
Use of steam pump at crib	299.60
Expenses to Cairo	300.00
Artesian well	341.54
Placing buoy for crib	141.95
Tug watching ways	810.93
Oil and candles	44.83
Sounding for crib	16.00
Use of boats	60.59
Locating crib	38.27
Nails and hardware	43.65
Repairing instruments	65.00
Carpenter work	74.00
Coal, lamps, and oil for crib	49.66
Labor	37.00
Field glass	23.00
Horse keeping	110.27
Lead	150.95
Use of capstan at crib	25.00
Chain	63.58
Surveyor's chain and tape measure	16.30
Advertising	28.88
Interest on payment to Dull & Gowen	97.77
Rope	49.12
Iron and labor at crib	130.89
Taking gauge at crib	22.50

1865	EXPENDITURES MINUS DEDUCTIONS ($)
Coal at inspectors' office	13.50
Miscellaneous	104.30
Labor at pumping works shop	316.51
Discount on 7 percent bonds	12,332.11
Deduction for expenses and labor refunded	<278.70>
TOTAL FOR 1865	230,220.08

1866 AND 1867	EXPENDITURES MINUS DEDUCTIONS ($)
Dull & Gowen, balance in full for all dues	94,562.89
Services of engineer	600.00
Services of draughtsman	1,058.32
Services of inspectors	7,789.75
Castings and iron work	332.16
Piles	201.80
Use of tugs	4,185.17
Artesian well	337.67
Labor	756.42
Cleaning tunnel	1,715.96
Horse keeping	140.52
Lead	52.68
Expenses of formal opening and celebration	979.18
Drifts from shafts	800.00
E. W. Smith, consulting engineer	600.00
Copies of plans	205.00
Connecting with well	180.86
Coal	363.00

1866 AND 1867	EXPENDITURES MINUS DEDUCTIONS ($)
Repairing lifeboat	63.81
Rope	61.73
Use of scow	58.00
Use of surveying equipment	50.00
Rubber boots	42.00
Lamps and oil	18.63
Brass figures	98.95
Repairing instruments	16.50
Securing lifeboat	15.00
Hose, pump, and pipe	13.60
Acid	11.25
Miscellaneous	70.50
Discount on 7 per cent bonds sold	2,353.24
Labor at shop	581.42
TOTAL FOR 1866 AND 1867	118,316.01
TOTAL FOR 1863 TO APRIL 1867	**457,844.95**

Sources: City of Chicago, *Third Annual Report of the Board of Public Works to the Common Council of the City of Chicago, April 1st, 1864* (Chicago: Jameson and Morse, 1864), 63–64, for 1863 expenditures; City of Chicago, *Fourth Annual Report of the Board of Public Works to the Common Council of the City of Chicago for the Municipal Fiscal Year Ending March 31st, 1865* (Chicago: George H. Fergus, 1865), 79, 80, for 1864 expenditures; City of Chicago, *Fifth Annual Report of the Board of Public Works to the Common Council of the City of Chicago for the Municipal Fiscal Year Ending March 31st, 1866* (Chicago: Jameson and Morse, 1866), appendixes 3–4, for 1865 expenditures; City of Chicago, *Sixth Annual Report of the Board of Public Works to the Common Council of the City of Chicago for the Municipal Fiscal Year Ending March 31st, 1867* (Chicago: Republican Book and Job Office, 1868), 55–56, for 1866 and 1867 expenditures.

Appendix E: Growth of the Chicago Waterworks, 1854–70

YEAR	AVG. DAILY PUMPAGE (GALS.)	AVG. DAILY PUMPAGE (PER CAPITA)	CAPACITY OF WORKS (GALS.)	NO. OF HOUSE SERVICES OR TAPS	NO. OF HYDRANTS	AMT. OF WATER PIPE (MI.)
1854	591,083	8.9	8,000,000	———*	123	30
1855	2,392,945	29.9	8,000,000	———*	158	43
1856	4,000,000	47.5	8,000,000	———*	169	52.4
1857	3,552,052	38.2	8,000,000	———*	182	58
1858	2,991,413	35.6	20,000,000	4,666	273	72.4
1859	3,877,119	41.2	20,000,000	5,511	388	85.1
1860	4,703,525	43	20,000,000	6,350	———**	91
1861	4,841,520	40.3	20,000,000	6,876	———**	95.3
1862	6,074,739	44.3	20,000,000	7,450	———**	104.9
1863	6,400,298	42.6	20,000,000	8,412	548	117.9
1864	6,913,259	42.8	20,000,000	9,855	606	127.3
1865	7,610,459	42.6	20,000,000	11,406	762	141.2
1866	8,681,536	43.3	20,000,000	13,634	790	156.5
1867	11,562,273	51.3	20,000,000	16,498	919	179.4
1868	14,724,999	58.4	38,000,000	20,915	1,070	208.6
1869	18,633,278	66.7	38,000,000	28,308	1,294	239.9
1870	21,766,260	70.9	38,000,000	35,318	1,552	272.4

* Number of house services or taps not reported in source material.
** Number of hydrants not reported in source material.
Source: John Ericson, *Report on the Water Supply System of Chicago in Past, Present and Future* (Chicago: Cameron, Amberg and Company, 1905), 38–39.

Appendix F: Chronology

1674	Louis Joliet suggests waterway connecting Lake Michigan and Mississippi River
1795	First official suggestion of canal connecting Lake Michigan and Mississippi River
1813	Ellis Sylvester Chesbrough born July 6, Baltimore County, Maryland
1816	The Potawatomi cede site of Chicago and Des Plaines and Illinois Valleys to United States
1818	Illinois becomes a state
1822	First act by Congress aiding Illinois canal
1823	First canal legislation by state; first surveys for canal
1825	Illinois and Michigan Canal Association incorporated
1834	Chicago Council authorizes digging of first public well
1836	Chicago Hydraulic Company incorporated
1836	Illinois and Michigan Canal begun
1848	Bridgeport pumping works constructed
1848	Illinois and Michigan Canal completed
1851	Chicago City Hydraulic Company incorporated; Board of Water Commissioners created
1855	Board of Sewerage Commissioners created; Chesbrough hired as chief engineer
1855	Chesbrough's plan for Chicago's sewerage system proposed
1856	Work on Chicago sewerage system begun
1858	Chesbrough issues report on European sewerage systems
1859	Chicago lake tunnel suggested for first time by Chesbrough
1861	Board of Public Works organized; Chesbrough appointed chief engineer
1863	Chesbrough appointed city engineer; plan for building Chicago lake tunnel approved
1864	Construction of lake tunnel and Two-Mile Crib begins
1865	Two-Mile Crib launched and placed
1866	Construction of lake tunnel completed
1867	Chicago lake tunnel officially opened
1867–70	Construction of new waterworks
1871	Illinois and Michigan Canal deepened; Ogden-Wentworth ditch constructed; Great Chicago Fire
1879	Chesbrough resigns from his employment with the City of Chicago
1886	Chesbrough dies August 18, Chicago
1936	Filling of the original lake tunnel and demolition of the Two-Mile Crib

NOTES

Introduction

1. Walter Nugent, "Demography: Chicago as a Modern World City," in Janice L. Reiff, Ann Durkin Keating, and James R. Grossman, eds., *Encyclopedia of Chicago* (Chicago: University of Chicago Press, 2004), http://www.encyclopedia.chicagohistory.org/pages/962.html, accessed August 1, 2016.

2. James O'Connell, *Chicago's Quest for Pure Water* (Washington, DC: Public Works Historical Society, 1976), 4.

3. Hydrologically, Lake Michigan and Lake Huron form one lake. Together, measured by surface area, they comprise the largest body of fresh water in the world.

4. John Hamilton, "The Epidemics of Chicago," *Bulletin of the Society of Medical History of Chicago* 1, no. 1 (1911, October), 80. The death toll in 1854 was 1,424 out of a population of 65,872.

5. For a description of the physical geography of the Chicago region, see David Solzman, *The Chicago River: An Illustrated History and Guide to the River and Its Waterways* (Chicago: University of Chicago Press, 2006), 9–24.

6. G. P. Brown, *Drainage Channel and Waterway: A History of the Effort to Secure an Effective and Harmless Method for the Disposal of the Sewage in the City of Chicago and to Create a Navigable Channel between Lake Michigan and the Mississippi River* (Chicago: R. R. Donnelly & Sons, 1894), 94–95.

7. Quoted in Brown, *Drainage Channel and Waterway*, 111–12.

8. Ibid., 116–17.

9. Charles H. Taylor, ed., *History of the Board of Trade of Chicago*, 3 vols. (Chicago: Robert O. Law, 1917), 1:135–36.

10. Roy Basler, ed., *The Collected Works of Abraham Lincoln*, 9 vols. (New Brunswick, NJ: Rutgers University Press, 1953), 1:483.

11. Donald Miller, *City of the Century: The Epic of Chicago and the Making of America* (New York: Simon & Schuster, 1996), 87–90, 96.

Chapter 1

1. The account that follows of the 1832 epidemic is taken from John Hamilton, "The Epidemics of Chicago," *Bulletin of the Society of Medical History of Chicago* 1, no. 1 (1911, October), 75–79, and G. F. Pyle, "The Diffusion of Cholera in the United States in the Nineteenth Century," *Geographical Analysis* 1 (1969, January), 60–65.

2. Hamilton, "The Epidemics of Chicago," 76.

3. Nelson Blake, *Water for the Cities: A History of the Urban Water Supply Problem in the United States* (Syracuse, NY: Syracuse University Press, 1956), 248 et seq.

4. G. P. Brown, *Drainage Channel and Waterway: A History of the Effort to Secure an Effective and Harmless Method for the Disposal of the Sewage in the City of Chicago and to Create a Navigable Channel between Lake Michigan and the Mississippi River* (Chicago: R. R. Donnelly & Sons, 1894), 27.
5. See Brown, *Drainage Channel and Waterway*, 26–28; Jennie Hall, *The Story of Chicago* (Chicago: Rand McNally, 1911), 155–60.
6. Joseph Kirkland, *The Story of Chicago* (Chicago: Dibble, 1892), 149.
7. City of Chicago, *Report of the Board of Health for the City of Chicago for 1867, 1868, and 1869, and a Sanitary History of the City of Chicago from 1833 to 1870* (Chicago: Lakeside Publishing and Printing, 1871), 11.
8. Brown, *Drainage Channel and Waterway*, 28; *Chicago Tribune*, December 28, 1850.
9. Brown, *Drainage Channel and Waterway*, 28.
10. Kirkland, *The Story of Chicago*, 155.
11. Edmund James, *The Charters of the City of Chicago* (Chicago: University of Chicago Press, 1898), 96.
12. Alfred T. Andreas, *History of Chicago, from the Earliest Period to the Present Time*, 3 vols. (Chicago: Alfred T. Andreas, 1884), 1:185.
13. Ibid.
14. *Chicago Daily American*, August 19, 1839.
15. Bessie Louise Pierce, *A History of Chicago from Town to City, 1948–1871*, 3 vols. (Chicago: University of Chicago Press, 1940), 1:352n214.
16. Hall, *The Story of Chicago*, 161; Daphne Christensen, ed., *Chicago Public Works: A History* (Chicago: Rand McNally, 1973), 66.
17. Brown, *Drainage Channel and Waterway*, 28–29; Andreas, *History of Chicago*, 1:185–86.
18. James, *The Charters of the City of Chicago*, 184; Weston Goodspeed and Daniel Healy, eds., *History of Cook County, Illinois: Being a General Survey of Cook County History Including a Condensed History of Chicago and Special Account of Districts Outside the City Limits from the Earliest Settlement to the Present Time*, 2 vols. (Chicago: Hammond Press, 1911), 1:167–68.
19. Andreas, *History of Chicago*, 1:186.
20. Elmer Riley, *The Development of Chicago and Vicinity as a Manufacturing Center Prior to 1880* (Chicago: McElroy, 1911), 68.
21. Pierce, *A History of Chicago from Town to City*, 1:352–53; Hall, *The Story of Chicago*, 162.
22. Hall, *The Story of Chicago*, 162.
23. William Bross, *History of Chicago* (Chicago: Jansen, McClurg & Company, 1876), 119.
24. Andreas, *History of Chicago*, 1:186.
25. Ibid.
26. Hamilton, "The Epidemics of Chicago," 80.
27. Ibid.
28. *Chicago Tribune*, December 28, 1850.
29. Ibid.
30. Ibid.
31. For an extended examination of how moving to a public water supply reflected an evolving "infrastructure of ideas," see Carl Smith, *City Water, City Life* (Chicago: University of Chicago Press, 2013).

Chapter 2

1. The account that follows of the first board of water commissioners is drawn from Alfred T. Andreas, *History of Chicago, from the Earliest Period to the Present Time*, 3 vols. (Chicago: Alfred T. Andreas, 1884–1886), 1:186–89, and City of Chicago, "Report Made to the Water Commissioners of the City of Chicago, September 26, 1851, on Supplying the City with Water," in Water Commissioners of the City of Chicago, *Report of the Water Commissioners of the City of Chicago, Made to the Common Council, December 8, 1851, Together with the Act of Incorporation and a Statement of the Financial Condition of the City, November 10, 1851* (Chicago: Seaton & Peck, 1851), 13 et seq.

2. City of Chicago, "Report Made to the Water Commissioners," 13.

3. Ibid., 16–17.

4. Ibid., 5, emphasis in original.

5. Ibid., 6.

6. Ibid., 6–7.

7. Ibid., 7–8.

8. Ibid., 16.

9. Andreas, *History of Chicago*, 1:187.

10. Ibid.; City of Chicago, *Mayor's Annual Message and Fifteenth Annual Report of the Department of Public Works to the City Council of the City of Chicago for the Fiscal Year Ending December 31, 1890* (Chicago: Cameron, Amberg and Company, 1891), 447.

11. City of Chicago, *The Waterworks System of the City of Chicago* (Chicago: Chicago Bureau of Public Efficiency, 1917), 91.

12. City of Chicago, *Report of the Board of Health for the City of Chicago for 1867, 1868, and 1869, and a Sanitary History of the City of Chicago from 1833 to 1870* (Chicago: Lakeside Publishing and Printing, 1871), 29.

13. *The Waterworks System of the City*, 90.

14. Andreas, *History of Chicago*, 1:188.

15. Ibid.; City of Chicago, *Mayor's Annual Message and Fifteenth Annual Report of the Department of Public Works*, 447.

16. William Bross, *History of Chicago* (Chicago: Jansen, McClurg & Company, 1876), 120.

17. *Chicago Tribune*, April 10, 1855.

18. J. M. Wing, *The Tunnels and Water System of Chicago: Under the Lake and Under the River* (Chicago: J. M. Wing & Company, 1874), 10–13.

19. Andreas, *History of Chicago*, 1:189.

20. Ibid.

21. Ibid. The semiannual reports made by the Board of Water Commissioners to the Common Council between 1858 and 1861 are another source of financial information.

22. City of Chicago, *Nineteenth Semi-Annual Report of the Board of Water Commissioners to the Common Council of the City of Chicago, January 1st, 1861* (Chicago: Jameson and Morse, 1861), 6.

23. Ibid., 29.

24. Ibid., 32, emphasis in original.

Chapter 3

1. *Chicago Tribune*, January 11, 1864.

2. Ibid.
3. John Hamilton, "The Epidemics of Chicago," *Bulletin of the Society of Medical History of Chicago* 1, no. 1 (1911, October), 80.
4. *Chicago Tribune*, June 22, 1854.
5. "An Act to Incorporate a Board of Sewerage Commissioners for the City of Chicago" (1855), para. 4, in Joseph E. Gary, comp., *Laws and Ordinances Governing the City of Chicago, January 1, 1866, with an Appendix, Containing the Former Legislation Relating to the City, and Notes of Decisions of the Supreme Court of Illinois Relating to Corporations* (Chicago: E. B. Myers and Chandler, 1866) 591–99.
6. George Soper, John Watson, and Arthur Martin, *A Report to the Chicago Real Estate Board on the Disposal of the Sewage and Protection of the Water Supply of Chicago, Illinois* (Chicago, 1915), 69.
7. City of Chicago, *Report and Plan of Sewerage for the City of Chicago, Illinois* (Chicago: Charles Scott, 1855).
8. Ibid., 5.
9. Ibid., 6.
10. City of Chicago, *Chicago Sewerage: Report of the Results of Examinations Made in Relation to Sewerage in Several European Cities, in the Winter of 1856–7* (Chicago: Charles Scott, 1858).
11. Ibid., 94.
12. Quoted in Josiah Seymour Currey, *Chicago: Its History and Its Builders*, 5 vols. (Chicago: S. J. Clarke, 1912), 2:191–92.
13. See Louis Cain, "Raising and Watering a City: Ellis Sylvester Chesbrough and Chicago's First Sanitation System," *Technology and Culture* (1972, July), 360–63.
14. The biographical account that follows is drawn from Anna Chesebrough Wildey, *Genealogy of the Descendants of William Chesebrough* (New York: Press of T. A. Wright, 1903); "Memoirs of Deceased Members: Ellis Sylvester Chesbrough," *Proceedings of the American Society of Civil Engineers* 15 (1889, November), 160–63; Benezette Williams, "E. S. Chesbrough: A Memoir," *Journal of the Association of Engineering Societies* 6, no. 1 (1886, November), 1–10; *Biographical Sketches of the Leading Men of Chicago, Photographically Illustrated by J. Carbutt* (Chicago: Wilson and St. Clair, 1868); and City of Chicago, *A Century of Progress in Waterworks: Chicago, 1833–1933* (Chicago: Bureau of Engineering, 1933). Of special note is the memoir by Williams, who was a colleague of Chesbrough. After receiving his degree in civil engineering from the University of Michigan, in 1869, Williams was employed in Chesbrough's private office. Williams left Chicago shortly thereafter to work first in the city engineer's office in Milwaukee and then on various railroad and mining projects in Illinois and Wisconsin. He returned to Chicago in July 1872 as assistant engineer under Chesbrough, where he worked in both the sewerage and waterworks departments. He eventually became superintendent of sewers and, upon Chesbrough's retirement, became Chicago's city engineer, holding that position until September 1879.
15. Daniel Calhoun, *American Civil Engineer: Origins and Conflict* (Cambridge: The Technology Press, Massachusetts Institute of Technology, 1960), 141–42.

16. As will become evident, this blending of practical, hands-on training from both traditions would come to define Chesbrough's professional life and accomplishments. For further discussion of these traditions and their impact on America's growth and development, see Nathan Rosenberg, *Technology and American Economic Growth* (White Plains, NY: Harper and Row, 1972), 24 et seq.

17. Nelson Blake, *Water for the Cities: A History of the Urban Water Supply Problem in the United States* (Syracuse, NY: Syracuse University Press, 1956), 211.

18. The biographical account that follows is drawn from "Memoirs of Deceased Members: John Bloomfield Jervis," *Proceedings of the American Society of Civil Engineers* 11 (1885, October), 109–16, and Gerard Koeppel, *Water for Gotham: A History* (Princeton, NJ: Princeton University Press, 2000), 185 et seq.

19. Richard Wood, *Stephen Harriman Long* (Glendale, CA: Arthur H. Clarke, 1966), 41; G. P. Brown, *Drainage Channel and Waterway: A History of the Effort to Secure an Effective and Harmless Method for the Disposal of the Sewage in the City of Chicago and to Create a Navigable Channel between Lake Michigan and the Mississippi River* (Chicago: R. R. Donnelly & Sons, 1894), 136; Cain, "Raising and Watering a City," 355n4.

Chapter 4

1. G. P. Brown, *Drainage Channel and Waterway: A History of the Effort to Secure an Effective and Harmless Method for the Disposal of the Sewage in the City of Chicago and to Create a Navigable Channel between Lake Michigan and the Mississippi River* (Chicago: R. R. Donnelly & Sons, 1894), 61.

2. City of Chicago, *Report and Plan of Sewerage for the City of Chicago, Illinois* (Chicago: Charles Scott, 1855), 6–7. Chesbrough had proposed a similar flushing system for the sewers themselves; see Langdon Pearse, "Chicago's Quest for Potable Water," *Water and Sewer Works* 102, no. 5 (1955, May), 183.

3. City of Chicago, *Chicago Sewerage: Report of the Results of Examinations Made in Relation to Sewerage in Several European Cities, in the Winter of 1856–7* (Chicago: Charles Scott, 1858), 93.

4. Ibid.

5. Ibid., 94.

6. City of Chicago, *Report of the Board of Sewerage Commissioners of the City of Chicago for the Half Year Ending June 30, 1860* (Chicago: Board of Sewerage Commissioners, 1860), 4.

7. Ibid., 9.

8. Ibid., 12.

9. Ibid., 13.

10. Ibid., 16–17.

11. City of Chicago, *Nineteenth Semi-Annual Report of the Board of Water Commissioners to the Common Council of the City of Chicago, January 1st, 1861* (Chicago: Jameson and Morse, 1861), 6.

12. Ibid., 6.

13. Ibid., 26.

14. Ibid., 27.

15. Ibid.

16. Ibid., 28.
17. Ibid., 6, emphasis in original.
18. Ibid., 32.
19. Ibid., 33.
20. Ibid.
21. Ibid., 34.
22. Ibid., 39.
23. Ibid., 39–40.
24. *Chicago Tribune*, February 21, 1862.
25. City of Chicago, *First Annual Report of the Board of Public Works to the Common Council of the City of Chicago, January 1st, 1862* (Chicago: S. P. Rounds, 1862), 5.
26. Ibid., 6–7.
27. Ibid., 41.
28. Ibid.
29. Ibid., 41–42.
30. Ibid., 43.
31. Ibid.
32. City of Chicago, *Second Annual Report of the Board of Public Works to the Common Council of the City of Chicago, April 1st, 1863* (Chicago: Tribune Book and Job Printing Company, 1863), 6.
33. Ibid., 38.
34. Ibid., 40.
35. Ibid.
36. Ibid., 42–43.
37. City of Chicago, *Eighth Annual Report of the Board of Public Works to the Common Council of the City of Chicago for the Municipal Fiscal Year Ending March 31st, 1869* (Chicago: Jameson and Morse, 1869), 94.
38. Ibid.
39. Ibid., 94–95.
40. Ibid., 95.
41. Ibid.
42. Ibid.
43. City of Chicago, *Second Annual Report of the Board of Public Works*, 49 (for cost analysis, see ibid., 45–50).
44. City of Chicago, *Eighth Annual Report of the Board of Public Works*, 96.
45. *Chicago Tribune*, June 13, 1863.
46. *Chicago Tribune*, June 17, 1863.
47. *Chicago Tribune*, August 4, 1863.
48. *Chicago Tribune*, August 11, 1863.
49. *Chicago Tribune*, August 17, 1863.
50. City of Chicago, *Eighth Annual Report of the Board of Public Works*, 96.
51. Ibid.
52. *Chicago Tribune*, September 11, 1863.
53. Ibid.
54. Ellis Chesbrough, letter to the Board of Public Works dated September 11, 1863, in possession of the Illinois Regional Archive Depository, Northeastern Illinois University.

55. *Chicago Tribune*, September 12, 1863.
56. City of Chicago, *Eighth Annual Report of the Board of Public Works*, 95–96.
57. *Chicago Tribune*, September 16, 1863.
58. Ibid.
59. *Chicago Tribune*, September 22, 1863.
60. See *Chicago Tribune*, October 6, 1863, for the account that follows of the meeting held on October 5.
61. *Laws and Ordinances Governing the City of Chicago, January 1, 1866, with an Appendix, Containing the Former Legislation Relating to the City, and Notes of Decisions of the Supreme Court of Illinois Relating to Corporations*, comp. Joseph Gary (Chicago: E. B. Myers and Chandler, 1866), 526–27.
62. *Chicago Tribune*, October 20, 1863.
63. *Chicago Tribune*, November 14, 1863.
64. Ibid.
65. *Chicago Tribune*, October 20, 1863.
66. *Chicago Tribune*, October 22, 1863.
67. *Chicago Tribune*, October 23, 1863.
68. *Chicago Tribune*, October 28, 1863.
69. City of Chicago, *Eighth Annual Report of the Board of Public Works*, 96.
70. *Chicago Tribune*, October 29, 1863.
71. *Chicago Tribune*, October 30, 1863.
72. *Chicago Tribune*, November 3, 1863.
73. *Chicago Tribune*, November 10, 1863.
74. *Chicago Tribune*, November 20, 1863.
75. City of Chicago, *Third Annual Report of the Board of Public Works to the Common Council of the City of Chicago, April 1st, 1864* (Chicago: Jameson and Morse, 1864), 119.
76. *Chicago Tribune*, November 24, 1863.
77. *Chicago Tribune*, December 4, 1863.
78. Ibid.
79. City of Chicago, *Third Annual Report of the Board of Public Works*, 122.

Chapter 5

1. *Chicago Tribune*, January 14, 1864; *Chicago Tribune*, January 25, 1864; *Chicago Tribune*, February 11, 1864; *Chicago Tribune*, March 12, 1864.
2. *Chicago Tribune*, March 18, 1864.
3. Ibid.
4. Chesbrough cited as an example of this pneumatic process the pile driving done in connection with the Savannah Bridge as described in M. Cezanne, "On the Theiss Bridge and Tubular Foundations," trans. J. Bennett, *Journal of the Franklin Institute, of the State of Pennsylvania, for the Promotion of the Mechanical Arts* (1861, October): 217–21.
5. *Chicago Tribune*, March 22, 1864; *Chicago Tribune*, April 7, 1864; *Chicago Tribune*, April, 9, 1864; *Chicago Tribune*, April 12, 1864.
6. *Chicago Tribune*, April 20, 1864; *Chicago Tribune*, April 25, 1864; *Chicago Tribune*, April 28, 1864.
7. *Chicago Tribune*, April 29, 1864.

8. *Chicago Tribune*, May 20, 1864.

9. City of Chicago, *Eighth Annual Report of the Board of Public Works to the Common Council of the City of Chicago for the Municipal Fiscal Year Ending March 31st, 1869* (Chicago: Jameson and Morse, 1869), 99.

10. J. M. Wing, *The Tunnels and Water System of Chicago: Under the Lake and Under the River* (Chicago: J. M. Wing & Company, 1874), 68.

11. Ibid.

12. City of Chicago, *Eighth Annual Report of the Board of Public Works*, 105.

13. Original payroll sheets are in the collection of the Chicago History Museum. A photograph of one of those sheets can be seen at Pamela Bannos, "The Earliest Years: The Lake Tunnel," Shifting Grounds, http://shifting-grounds.net/pre-history/tunnel.html, accessed August 1, 2016.

14. City of Chicago, *Eighth Annual Report of the Board of Public Works*, 99.

15. Ibid., 99–100. These encounters with large boulders during the tunneling opened up an interesting question. In the mid-1800s, there was still considerable scientific debate on the nature of glaciers and their relationship to a possible ice age. Indeed, there was widespread disagreement over the question of whether the Ice Age proper had ever actually existed. A leading proponent of the theory that the Ice Age had existed was Louis Agassiz, and the discovery of boulders at a depth of 75 feet—boulders marked by grooves and parallel lines indicative of glacial movement—was seen as evidence in support of his theory; see *Chicago Tribune*, May 4, 1864. In July 1864, Chesbrough was invited to make a presentation at the Chicago Academy of Sciences, to which he donated "certain boulders . . . [that] presented strong indications of the actions of glaciers" as well as "soft slate boulders containing pyrites of iron, and various other interesting specimens from the same excavation, with numerous descriptions of fossils, iron pyrites and crystalized limestone taken from an adjoining quarry," and he also presented a sectional drawing of the progress of the lake tunnel up to that point, which, together with the boulders, "excited much curiosity" on the part of those in attendance; see *Chicago Tribune*, July 14, 1864. Five years later he wrote, with understandable pride, "The Academy of Sciences of this city had representatives, Drs. Jewell and Andrews, who made numerous observations relative to the geological developments by the excavations for the tunnel. The result of these has already been published in different forms, in Silliman's and other scientific journals"; see City of Chicago, *Eighth Annual Report of the Board of Public Works*, 119. Unfortunately, the items donated by Chesbrough were lost during the Great Chicago Fire of 1871, but the surviving records of the Chicago Academy of Sciences do contain an entry recording a donation by Chesbrough of "2 specimens of wood found 33 feet below the surface in shaft for the new Lake Tunnel" (archive entry courtesy of Dawn Roberts, collection manager of the Chicago Academy of Sciences).

16. *Chicago Tribune*, June 6, 1864.

17. City of Chicago, *Eighth Annual Report of the Board of Public Works*, 101.

18. Wing, *The Tunnels and Water System of Chicago*, 58.

19. City of Chicago, *Eighth Annual Report of the Board of Public Works*, 102.

20. Ibid., 103; City of Chicago, *Fourth Annual Report of the Board of Public Works to the Common Council of the City of Chicago for the Municipal Fiscal Year Ending March 31st, 1865* (Chicago: George H. Fergus, 1865), 55n.

21. *Chicago Tribune*, August 24, 1864.
22. Ibid.
23. Handwritten invitation from Dull & Gowen in possession of the Illinois Regional Archives Depository, Northeastern Illinois University.
24. *Chicago Tribune*, December 2, 1864.
25. Ibid.
26. Ibid.

Chapter 6

1. *New York Times*, July 30, 1865.
2. *Chicago Tribune*, July 25, 1865.
3. The description that follows of the crib's structure and construction relies primarily on Chesbrough's account in City of Chicago, *Eighth Annual Report of the Board of Public Works to the Common Council of the City of Chicago for the Municipal Fiscal Year Ending March 31st, 1869* (Chicago: Jameson and Morse, 1869), 106–8. Other sources include J. M. Wing, *The Tunnels and Water System of Chicago: Under the Lake and Under the River* (Chicago: J. M. Wing & Company, 1874), 39–40; *Chicago Tribune*, July 25, 1865; and *New York Times*, July 30, 1865. However, it should be noted that the latter sources differ, sometimes significantly, from Chesbrough's account. Deference is here given to Chesbrough, who provided the details of the crib's construction in 1869 and presumably was more accurate in his description than the contemporaneous newspaper accounts that formed the basis for Wing's account.
4. City of Chicago, *Eighth Annual Report of the Board of Public Works*, 106–8.
5. Wing, *The Tunnels and Water System of Chicago*, 43.
6. *Chicago Tribune*, July 25, 1865.
7. City of Chicago, *Eighth Annual Report of the Board of Public Works*, 110.
8. Ibid., 111–12.
9. *Chicago Tribune*, November 30, 1865.
10. In one of the more unseemly turns of events associated with the construction of the lake tunnel, Bramhall eventually had to sue the contractors for the work he had done on launching the crib, lowering the cylinders, and creating the plans for installing the mooring screws. The contractors offered to settle the suit for $200, but Bramhall refused. At the ensuing trial, testimony was given that Bramhall, after telling James Gowen of his intention to charge for his plans, was told by Gowen that he would be well paid for the plans once the work was done, a promise that Gowen allegedly reiterated when he reportedly assured Bramhall that he would be paid for the plans if the contractors were paid by the city, and that he would be paid something for the plans in any event. The jury found for Bramhall and awarded him $650. Dull & Gowen appealed, claiming that all Bramhall's work had been covered under the initial contract between the contractors and Bramhall. The Illinois Supreme Court affirmed the lower court's judgment, noting that Bramhall's plans had exceeded any work that he had been hired to do as superintendent of carpentry. The court went on to note that Bramhall's plans "were acknowledged to be very ingenious, and quite successful, giving evidence of such inventive genius as in most countries would have been attended by high honor and great reward." The court called the lower court's judgment "a pittance," given the extra work that Bramhall had provided, and ordered Dull

& Gowen to pay the judgment; see Norman Freeman, rept., *Reports of Cases at Law and in Chancery Argued and Determined in the Supreme Court of Illinois* (Springfield, IL: Journal Company, 1870), 364–68.

11. City of Chicago, *Eighth Annual Report of the Board of Public Works*, 110.
12. Ibid.
13. Ibid.
14. Ibid.
15. *Chicago Tribune*, February 24, 1865.
16. *Chicago Tribune,* March 7, 1865.
17. *Chicago Tribune,* March 21, 1865.
18. *Chicago Tribune,* April 2, 1865.
19. *Chicago Tribune,* April 27, 1865.
20. *Chicago Tribune,* January 23, 1866.
21. *Chicago Tribune*, February 2, 1866.
22. *Chicago Tribune*, February 13, 1866.
23. *Chicago Tribune,* February 20, 1866.
24. *Chicago Tribune,* February 21, 1866.

Chapter 7

1. *Chicago Tribune*, March 22, 1866.
2. City of Chicago, *Eighth Annual Report of the Board of Public Works to the Common Council of the City of Chicago for the Municipal Fiscal Year Ending March 31st, 1869* (Chicago: Jameson and Morse, 1869), 113.
3. *Chicago Tribune*, January 11, 1866.
4. J. M. Wing, *The Tunnels and Water System of Chicago: Under the Lake and Under the River* (Chicago: J. M. Wing & Company, 1874), 38.
5. Ibid., 58–60.
6. *Chicago Tribune*, May 7, 1866.
7. *Chicago Tribune*, October 8, 1866.
8. City of Chicago, *Eighth Annual Report of the Board of Public Works*, 119.
9. *Chicago Tribune*, November 25, 1866.
10. *Chicago Tribune*, November 27, 1866; *Chicago Tribune*, November 29, 1866.
11. *Chicago Tribune*, December 18, 1866.
12. *Chicago Tribune*, November 28, 1866.
13. *Chicago Tribune*, December 2, 1866; Wing, *The Tunnels and Water System of Chicago*, 78–79.
14. *Chicago Tribune*, November 28, 1866.
15. *Chicago Tribune*, December 7, 1866. The *Tribune*'s account of the celebration and ceremonies varies in some details—most notably with respect to the manner of transport for the parties, and the language of the second speech by Mayor Rice—from the account given in Wing, *The Tunnels and Water System of Chicago*, 82–86; for consistency, however, the more contemporaneous account is used here.
16. *Chicago Tribune*, December 7, 1866.
17. Ibid.
18. Ibid.
19. Ibid.

20. Ibid.
21. Ibid.
22. Ibid.
23. Ibid.
24. *Chicago Tribune*, December 13, 1866; *Chicago Tribune*, December 19, 1866.
25. City of Chicago, *Fifth Annual Report of the Board of Public Works to the Common Council of the City of Chicago for the Municipal Fiscal Year Ending March 31st, 1866* (Chicago: Jameson and Morse, 1866), 2.
26. City of Chicago, *Eighth Annual Report of the Board of Public Works*, 66–70.
27. City of Chicago, *Fifth Annual Report of the Board of Public Works*, 5–7.
28. City of Chicago, *Eighth Annual Report of the Board of Public Works*, 113–14.
29. City of Chicago, *Sixth Annual Report of the Board of Public Works to the Common Council of the City of Chicago for the Municipal Fiscal Year Ending March 31st, 1867* (Chicago: Republican Book and Job Office, 1868), 36.
30. *Chicago Tribune*, March 8, 1867; *Chicago Tribune*, March 9, 1867.
31. *Chicago Tribune*, March 25, 1867.
32. Ibid.
33. Ibid.
34. City of Chicago, *Eighth Annual Report of the Board of Public Works*, 115.
35. Ibid.
36. *Chicago Tribune*, March 23, 1867.
37. Original program in possession of the Chicago Historical Museum.
38. *Chicago Tribune*, March 25, 1867.
39. *Chicago Tribune*, March 26, 1867.
40. Ibid.
41. Ibid.
42. Ibid.
43. Ibid.
44. The Chicago Citizens' Testimonial and a letter of thanks from Chesbrough dated November 29, 1867, in possession of the Special Collections at the Harold Washington Library, Chicago.

Chapter 8

1. City of Chicago, *Sixth Annual Report of the Board of Public Works to the Common Council of the City of Chicago for the Municipal Fiscal Year Ending March 31st, 1867* (Chicago: Republican Book and Job Office, 1868), 35.
2. City of Chicago, *Seventh Annual Report of the Board of Public Works to the Common Council of the City of Chicago for the Municipal Fiscal Year Ending March 31st, 1868* (Chicago: Jameson and Morse, 1869), 45.
3. Ibid., 59.
4. Ibid., 43.
5. Ibid., 60.
6. City of Chicago, *Eighth Annual Report of the Board of Public Works to the Common Council of the City of Chicago for the Municipal Fiscal Year Ending March 31st, 1869* (Chicago: Jameson and Morse, 1869), 71–72.

7. "Throwback Thursday: Chicago Water Tower Edition," Chicago Architecture, March 5, 2015, http://www.chicagoarchitecture.org/2015/03/05/throwback-thursday-chicago-water-tower-edition. The National Park Service surveyed the water tower in connection with the service's Historic American Buildings Survey, survey no. HABS IL-1041, Paul Berry, delineator.

8. *Chicago Tribune*, September 11, 1869.

9. City of Chicago, *Ninth Annual Report of the Board of Public Works to the Common Council of the City of Chicago for the Municipal Fiscal Year Ending March 31st, 1870* (Chicago: Lakeside Publishing and Printing, 1870), 84.

10. Ibid., 86.

Chapter 9

1. *Chicago Tribune*, November 28, 1866.

2. William Blake, *Civil Engineering and Public Works, Paris Universal Exposition 1867: Reports of the United States Commissioners* (Washington, DC: Government Printing Office, 1870), 17.

3. Ibid., 16.

4. City of Chicago, *Eighth Annual Report of the Board of Public Works to the Common Council of the City of Chicago for the Municipal Fiscal Year Ending March 31st, 1869* (Chicago: Jameson and Morse, 1869), 117–18.

5. City of Chicago, *Eleventh Annual Report of the Board of Public Works to the Common Council of the City of Chicago for the Municipal Fiscal Year Ending March 31st, 1872* (Chicago: D. & C. H. Blakely, 1872), 49.

6. City of Chicago, *Tenth Annual Report of the Board of Public Works to the Common Council of the City of Chicago for the Municipal Fiscal Year Ending March 31st, 1871* (Cambridge, MA: Welch, Bigelow, & Company, 1871), 62.

7. Undated letter from Chesbrough, copy in possession of the Chicago Historical Museum.

8. City of Chicago, *Eleventh Annual Report of the Board of Public Works*, 3–4.

9. Ibid., 50.

10. Ibid., 50–51.

11. Ibid., 5–6.

12. Ibid., 52–53.

13. City of Chicago, *Twelfth Annual Report of the Board of Public Works to the Common Council of the City of Chicago for the Municipal Fiscal Year Ending March 31st, 1873* (Chicago: National Printing Company, 1873), 47.

Epilogue

1. In 1858, average daily per capita consumption was 32.8 gallons; by 1870, it was 72.8 gallons. See City of Chicago, *Tenth Annual Report of the Board of Public Works to the Common Council of the City of Chicago for the Municipal Fiscal Year Ending December 31st, 1885* (Chicago: Cameron, Amberg and Company, 1886), 66–67.

2. *Chicago Tribune*, May 16, 1879.

3. Cliff Schexnayder, *Builders of the Hoosac Tunnel* (Portsmouth, NH: Peter E. Randall, 2015), 298–99.

4. Henry Drinker, *Tunneling, Explosive Compounds, and Rock Drills* (New York: John Wiley & Sons, 1893), 939–55.

5. *Chicago Tribune*, June 9, 1903.

6. The Lawrence Avenue lake tunnel and crib, although not associated with the city's water supply, were completed in 1908 as part of the city's intercepting sewer system, and the lake tunnel was approximately 1,500 feet long; see City of Chicago, *Mayor's Annual Message and Thirty-Third Annual Report of the Department of Public Works to the City Council of the City of Chicago for the Year Ending December 31, 1908* (Chicago: Cameron, Amberg and Company, 1909), 11, and City of Chicago, *Mayor's Annual Message and Thirty-First Annual Report of the Department of Public Works to the City Council of the City of Chicago for the Year Ending December 31, 1906* (Chicago: W. J. Hartman Company, 1907), 183. In 1932, the crib was abandoned and subsumed by the sand fill for the northward expansion of Lincoln Park; see Michael J. Chrzastowski, *"Make No Little Plans": Field Trip Guidebook for the American Shore and Beach Preservation Association 2008 National Conference* (Champaign: Illinois Geological Survey, 2008), 30.

7. The upcoming demolition of the crib was announced in the *Chicago Tribune*, September 19, 1936.

8. City of Chicago, *Sixty-First Annual Report of the Department of Public Works for the Year Ending December 31, 1936* (Chicago: Fred J. Rigley, 1937), 179.

BIBLIOGRAPHY

Books and Pamphlets

Andreas, Alfred T. *History of Chicago, from the Earliest Period to the Present Time*. 3 vols. Chicago: Alfred T. Andreas, 1884–1886.

Basler, Roy, ed. *The Collected Works of Abraham Lincoln*. 9 vols. New Brunswick, NJ: Rutgers University Press, 1953.

Biographical Sketches of the Leading Men of Chicago, Photographically Illustrated by J. Carbutt. Chicago: Wilson & St. Clair, 1868.

Blake, Nelson. *Water for the Cities: A History of the Urban Water Supply Problem in the United States*. Syracuse, NY: Syracuse University Press, 1956.

Blake, William. *Civil Engineering and Public Works, Paris Universal Exposition 1867: Reports of the United States Commissioners*. Washington, DC: Government Printing Office, 1870.

Bross, William. *History of Chicago*. Chicago: Jansen, McClurg & Company, 1876.

Brown, G. P. *Drainage Channel and Waterway: A History of the Effort to Secure an Effective and Harmless Method for the Disposal of the Sewage in the City of Chicago and to Create a Navigable Channel between Lake Michigan and the Mississippi River*. Chicago: R. R. Donnelly & Sons, 1894.

Cain, Louis. *Sanitation Strategy for a Lakefront Metropolis: The Case of Chicago*. DeKalb: Northern Illinois University Press, 1978.

Calhoun, Daniel. *American Civil Engineer: Origins and Conflict*. Cambridge: The Technology Press, Massachusetts Institute of Technology, 1960.

Christensen, Daphne, ed. *Chicago Public Works: A History*. Chicago: Rand McNally, 1973.

Chrzastowski, Michael J. *"Make No Little Plans": Field Trip Guidebook for the American Shore and Beach Preservation Association 2008 National Conference*. Champaign: Illinois Geological Survey, 2008.

Civil News Service of Illinois. *A Pictorial Story of Chicago's Public Works: Engineering Wonders of the World, Featuring Chicago's Charter Jubilee, 1837–1937*. Chicago: Civil News Service of Illinois, 1937.

Currey, Josiah Seymour. *Chicago: Its History and Its Builders*. 5 vols. Chicago: S. J. Clarke, 1912.

Drinker, Henry. *Tunneling, Explosive Compounds, and Rock Drills*. New York: John Wiley & Sons, 1893.

Ericson, John. *Report on the Water Supply System of Chicago in Past, Present and Future*. Chicago: Cameron, Amberg and Company, 1905.

Goodspeed, Weston, and Daniel Healy, eds. *History of Cook County, Illinois: Being a General Survey of Cook County History Including a Condensed History of Chicago and Special Account of Districts Outside the City Limits from the Earliest Settlement to the Present Time*. 2 vols. Chicago: Hammond Press, 1911.

Hall, Jennie. *The Story of Chicago*. Chicago: Rand McNally, 1911.

Hoyt, Homer. *One Hundred Years of Land Values in Chicago: The Relationship of the Growth of Chicago to the Rise of Its Land Values, 1830–1933*. Chicago: University of Chicago Press, 1933.

James, Edmund. *The Charters of the City of Chicago*. Chicago: University of Chicago Press, 1898.

Kirkland, Joseph. *The Story of Chicago*. Chicago: Dibble, 1892.

Koeppel, Gerard. *Water for Gotham: A History*. Princeton, NJ: Princeton University Press, 2000.

Miller, Donald. *City of the Century: The Epic of Chicago and the Making of America*. New York: Simon & Schuster, 1996.

Mitchell, Samuel Augustus. *Mitchell's New General Atlas*. Philadelphia, PA: S. A. Mitchell Jr., 1870.

Nugent, Walter. "Demography: Chicago as a Modern World City." In Janice L. Reiff, Ann Durkin Keating, and James R. Grossman, eds., *Encyclopedia of Chicago*. Chicago: University of Chicago Press, 2004. Available at http://www.encyclopedia.chicagohistory.org/pages/962.html, accessed August 1, 2016.

O'Connell, James. *Chicago's Quest for Pure Water*. Washington, DC: Public Works Historical Society, 1976.

Pierce, Bessie Louise. *A History of Chicago from Town to City, 1948–1871*. 3 vols. Chicago: University of Chicago Press, 1940.

Riley, Elmer. *The Development of Chicago and Vicinity as a Manufacturing Center prior to 1880*. Chicago: McElroy, 1911.

Rosenberg, Nathan. *Technology and American Economic Growth*. White Plains, NY: Harper and Row, 1972.

Schexnayder, Cliff. *Builders of the Hoosac Tunnel*. Portsmouth, NH: Peter E. Randall, 2015.

Smith, Carl. *City Water, City Life*. Chicago: University of Chicago Press, 2013.

Solzman, David. *The Chicago River: An Illustrated History and Guide to the River and Its Waterways*. Chicago: University of Chicago Press, 2006.

Soper, George, John Watson, and Arthur Martin. *A Report to the Chicago Real Estate Board on the Disposal of the Sewage and Protection of the Water Supply of Chicago, Illinois*. Chicago, 1915.

Taylor, Charles H., ed. *History of the Board of Trade of Chicago*. 3 vols. Chicago: Robert O. Law, 1917.

Wildey, Anna Chesebrough. *Genealogy of the Descendants of William Chesebrough*. New York: Press of T. A. Wright, 1903.

Wing, J. M. *The Tunnels and Water System of Chicago: Under the Lake and Under the River*. Chicago: J. M. Wing & Company, 1874.

Wood, Richard. *Stephen Harriman Long*. Glendale, CA: Arthur H. Clarke Company, 1966.

Periodicals and Websites

Bannos, Pamela. "Block 21 and Chicago's MCA." Shifting Grounds, http://shifting-grounds.net, accessed August 1, 2016.

———. "The Earliest Years: The Lake Tunnel." Shifting Grounds, http://shifting-grounds.net/pre-history/tunnel.html, accessed August 1, 2016.

Cain, Louis. "Raising and Watering a City: Ellis Sylvester Chesbrough and Chicago's First Sanitation System." *Technology and Culture* (1972, July): 353–72.

Cezanne, M. "On the Theiss Bridge and Tubular Foundations," trans. J. Bennett. *Journal of the Franklin Institute, of the State of Pennsylvania, for the Promotion of the Mechanical Arts* (1861, Oct.): 217–21.

Hamilton, John. "The Epidemics of Chicago." *Bulletin of the Society of Medical History of Chicago* 1, no. 1 (1911, Oct.): 73–86.

"Memoirs of Deceased Members: Ellis Sylvester Chesbrough." *Proceedings of the American Society of Civil Engineers* 15 (1889, Nov.): 160–63.

"Memoirs of Deceased Members: John Bloomfield Jervis." *Proceedings of the American Society of Civil Engineers* 11 (1885, Oct.): 109–16.

Pearse, Langdon. "Chicago's Quest for Potable Water." *Water and Sewer Works* 102, no. 5 (1955, May): 183–91.

Pyle, G. F. "The Diffusion of Cholera in the United States in the Nineteenth Century." *Geographical Analysis* 1 (1969, Jan.): 59–75.

Williams, Benezette. "E. S. Chesbrough: A Memoir." *Journal of the Association of Engineering Societies* 6, no. 1 (1886, Nov.): 1–10.

Public Records and Documents

State of Illinois and City of Chicago: Court Cases, Laws, and Ordinances

Freeman, Norman, rept. *Reports of Cases at Law and in Chancery Argued and Determined in the Supreme Court of Illinois*. Vol. 49. Springfield, IL: Journal Company, 1870.

Gary, Joseph E., comp. *Laws and Ordinances Governing the City of Chicago, January 1, 1866, with an Appendix, Containing the Former Legislation Relating to the City, and Notes of Decisions of the Supreme Court of Illinois Relating to Corporations*. Chicago: E. B. Myers and Chandler, 1866.

City of Chicago: Public Health

City of Chicago. *Report of the Board of Health for the City of Chicago for 1867, 1868, and 1869, and a Sanitary History of the City of Chicago from 1833 to 1870*. Chicago: Lakeside Publishing and Printing, 1871.

City of Chicago: Public Works

City of Chicago. *First Annual Report of the Board of Public Works to the Common Council of the City of Chicago, January 1st, 1862*. Chicago: S. P. Rounds, 1862.

City of Chicago. *First Annual Report of the Department of Public Works to the City Council of the City of Chicago for the Fiscal Year Ending December 31, 1876*. Chicago: Clark & Edwards, 1877.

City of Chicago. *Second Annual Report of the Board of Public Works to the Common Council of the City of Chicago, April 1st, 1863*. Chicago: Tribune Book and Job Printing Company, 1863.

City of Chicago. *Third Annual Report of the Board of Public Works to the Common Council of the City of Chicago, April 1st, 1864*. Chicago: Jameson and Morse, 1864.

City of Chicago. *Fourth Annual Report of the Board of Public Works to the Common Council of the City of Chicago for the Municipal Fiscal Year Ending March 31st, 1865*. Chicago: George H. Fergus, 1865.

City of Chicago. *Fifth Annual Report of the Board of Public Works to the Common Council of the City of Chicago for the Municipal Fiscal Year Ending March 31st, 1866*. Chicago: Jameson and Morse, 1866.

City of Chicago. *Sixth Annual Report of the Board of Public Works to the Common Council of the City of Chicago for the Municipal Fiscal Year Ending March 31st, 1867*. Chicago: Republican Book and Job Office, 1868.

City of Chicago. *Seventh Annual Report of the Board of Public Works to the Common Council of the City of Chicago for the Municipal Fiscal Year Ending March 31st, 1868*. Chicago: Jameson and Morse, 1869.

City of Chicago. *Eighth Annual Report of the Board of Public Works to the Common Council of the City of Chicago for the Municipal Fiscal Year Ending March 31st, 1869*. Chicago: Jameson and Morse, 1869.

City of Chicago. *Ninth Annual Report of the Board of Public Works to the Common Council of the City of Chicago for the Municipal Fiscal Year Ending March 31st, 1870*. Chicago: Lakeside Publishing and Printing, 1870.

City of Chicago. *Tenth Annual Report of the Board of Public Works to the Common Council of the City of Chicago for the Municipal Fiscal Year Ending December 31st, 1885*. Chicago: Cameron, Amberg and Company, 1886.

City of Chicago. *Eleventh Annual Report of the Board of Public Works to the Common Council of the City of Chicago for the Municipal Fiscal Year Ending March 31st, 1872*. Chicago: D. & C. H. Blakely, 1872.

City of Chicago. *Twelfth Annual Report of the Board of Public Works to the Common Council of the City of Chicago for the Municipal Fiscal Year Ending March 31st, 1873*. Chicago: National Printing Company, 1873.

City of Chicago. *Mayor's Annual Message and Fifteenth Annual Report of the Department of Public Works to the City Council of the City of Chicago for the Fiscal Year Ending December 31, 1890*. Chicago: Cameron, Amberg and Company, 1891.

City of Chicago. *Mayor's Annual Message and Thirty-First Annual Report of the Department of Public Works to the City Council of the City of Chicago for the Year Ending December 31, 1906*. Chicago: W. J. Hartman Company, 1907.

City of Chicago. *Mayor's Annual Message and Thirty-Third Annual Report of the Department of Public Works to the City Council of the City of Chicago for the Year Ending December 31, 1908*. Chicago: Cameron, Amberg and Company, 1909.

City of Chicago. *Sixty-First Annual Report of the Department of Public Works for the Year Ending December 31, 1936*. Chicago: Fred J. Rigley, 1937.

City of Chicago: Sewerage

City of Chicago. *Chicago Sewerage: Report of the Results of Examinations Made in Relation to Sewerage in Several European Cities, in the Winter of 1856–7*. Chicago: Charles Scott, 1858.

City of Chicago. *Report of the Board of Sewerage Commissioners of the City of Chicago for the Half Year Ending June 30, 1860*. Chicago: Board of Sewerage Commissioners, 1860.

City of Chicago. *Report and Plan of Sewerage for the City of Chicago, Illinois.* Chicago: Charles Scott, 1855.

City of Chicago: Water

City of Chicago. *A Century of Progress in Waterworks: Chicago, 1833–1933.* Chicago: Bureau of Engineering, 1933.

City of Chicago. *Nineteenth Semi-Annual Report of the Board of Water Commissioners to the Common Council of the City of Chicago, January 1st, 1861.* Chicago: Jameson and Morse, 1861.

City of Chicago. "Report Made to the Water Commissioners of the City of Chicago, September 26, 1851, on Supplying the City with Water." In Water Commissioners of the City of Chicago, *Report of the Water Commissioners of the City of Chicago, Made to the Common Council, December 8, 1851, Together with the Act of Incorporation and a Statement of the Financial Condition of the City, November 10, 1851*, 3–27. Chicago: Seaton & Peck, 1851.

City of Chicago. *The Waterworks System of the City of Chicago.* Chicago: Chicago Bureau of Public Efficiency, 1917.

INDEX

Page numbers in *italics* refer to illustrations.

Second to None: Chicago Stories celebrates the authenticity of a city brimming with rich narratives and untold histories. Spotlighting original, unique, and rarely explored stories, Second to None unveils a new and significant layer to Chicago's big-shouldered literary landscape.

Harvey Young, series editor